Shells
from Cape Cod to Cape May
WITH SPECIAL REFERENCE TO THE NEW YORK CITY AREA

Shells
from Cape Cod to Cape May
WITH SPECIAL REFERENCE TO THE NEW YORK CITY AREA

(former title: Shells of the New York City Area)

MORRIS K. JACOBSON
*Associate in Malacology, The American Museum
of Natural History*

and

WILLIAM K. EMERSON
*Curator of Living Invertebrates, The American Museum
of Natural History*

Drawings by **ANTHONY D'ATTILIO**
San Diego Natural History Museum

Foreword by **R. TUCKER ABBOTT**
*du Pont Chair of Malacology, Delaware Museum
of Natural History*

Dover Publications, Inc.
New York

Copyright © 1961, 1971 by Morris K. Jacobson and William K. Emerson.
All rights reserved under Pan American and International Copyright Conventions.

Published in Canada by General Publishing Company, Ltd., 30 Lesmill Road, Don Mills, Toronto, Ontario.
Published in the United Kingdom by Constable and Company, Ltd., 10 Orange Street, London WC 2.

This Dover edition, first published in 1971, is a revised republication of the work orginally published by Argonaut Books, Inc., Larchmont, N.Y., in 1961 with the title *Shells of the New York City Area*. The authors have written a new Preface, have made several alterations in the text, and have supplied a list of new nomenclature, additions to the Bibliography and a Supplement of Additional Species (with new drawings by Anthony D'Attilio), all prepared specially for the present edition.

International Standard Book Number: 0-486-21402-8
Library of Congress Catalog Card Number: 71-164734

Manufactured in the United States of America
Dover Publications, Inc.
180 Varick Street
New York, N.Y. 10014

Preface
to the Dover Edition

In the ten years that have elapsed since the publication of the original edition of this book, new information of various kinds has come to light. For this reason, we have added a supplement of fifteen additional species (p. 137) to the number treated earlier. These include three land snails, three fresh-water snails, five marine gastropods and four marine pelecypods. These species are now described and several are illustrated with drawings by Anthony D'Attilio.

Recent advances in scientific knowledge require changes in the nomenclature of some of the local species. These changes are indicated in the list following this Preface. We also have added to the Bibliography titles of recently published books and articles pertaining to the mollusks of this region. Some alterations have been made in the text, and some footnotes added.

We wish to thank again the numerous collectors who have added to our knowledge of the local fauna.

New York City
April 1, 1971

MORRIS K. JACOBSON
WILLIAM K. EMERSON

LIST OF CHANGES IN NOMENCLATURE

Page 10, line 28: *"Succinea avara* Say" should read *"Catinella vermeta* Say."

Page 26, line 2: *"Lymnaea palustris elodes* Say" should read *"Lymnaea palustris* Müller."

Page 35, line 8: *"Viviparus contectoides* Binney" should read *"Viviparus georgianus* Lea."

Page 39, line 1: "*Elliptio complanatus* Dillwyn" should read "*Elliptio complanata* Dillwyn."

Page 40, line 16: "*Margaritana margaritifera* Linné" should read "*Margaritifera margaritifera* Linné."

Page 56, line 6: "*Thais lapillus* Linné" should read "*Nucella lapillus* Linné."

Page 58, line 16: "*Nassarius obsoletus* Say" should read "*Ilyanassa obsoleta* Say." The subgeneric name *Ilyanassa* is now afforded full generic rank.

Page 61, line 6: "*Colus islandicus* Gmelin" should read "*Colus islandica* Gmelin."

Page 74, line 28: "*Anadara pexata* Say" should read "*Anadara ovalis* Bruguière."

Naturally, these changes apply wherever the above shells are named in the book.

Foreword

New Yorkers have an undeserved reputation of being subway moles living in a greenless world of night lights and smog. This idea is preposterous for two obvious reasons. Firstly, there is as much natural history to be observed in the parks, backyards and nearby beaches of New York City as there is in almost any other part of our country. Within a fifty mile radius of Times Square there live over 140 species of land, fresh-water or marine shells, not to mention the other thousands of species of plants and animals. Secondly, New York City is especially noted for its magnificent museum of natural history and for the large number of naturalists among its citizens.

The New York Shell Club has for years nurtured the interests of a growing, enthusiastic band of shell collectors, and its former President, Morris K. Jacobson, has long realized the need for a handbook for the local conchologist. Dr. William K. Emerson of The American Museum of Natural History and Anthony D'Attilio, both club members, have added their own valuable talents to this useful and readable guide.

For some unexplained reason, shell collecting is the most insidious and habit-forming of the natural history hobbies. Most people who begin assembling a small, casual collection of shells eventually become severely shell-shocked and join the happy band of intense conchologists. Perhaps there is some Freudian explanation, for Neanderthal man collected shells and very young children still show a natural affinity towards these wonderful "artifacts" of nature. Some collectors of shell rarities have been accused of base greed, while others, on the other hand, have amassed good collections for the sheer scientific satisfaction of learning new conchological facts. If you live in one of the cities of northeastern United States, or are bored with the set patterns of Suburbia, you will be glad that you opened the covers of this entertaining shell book. Read it, and start collecting shells!

R. TUCKER ABBOTT

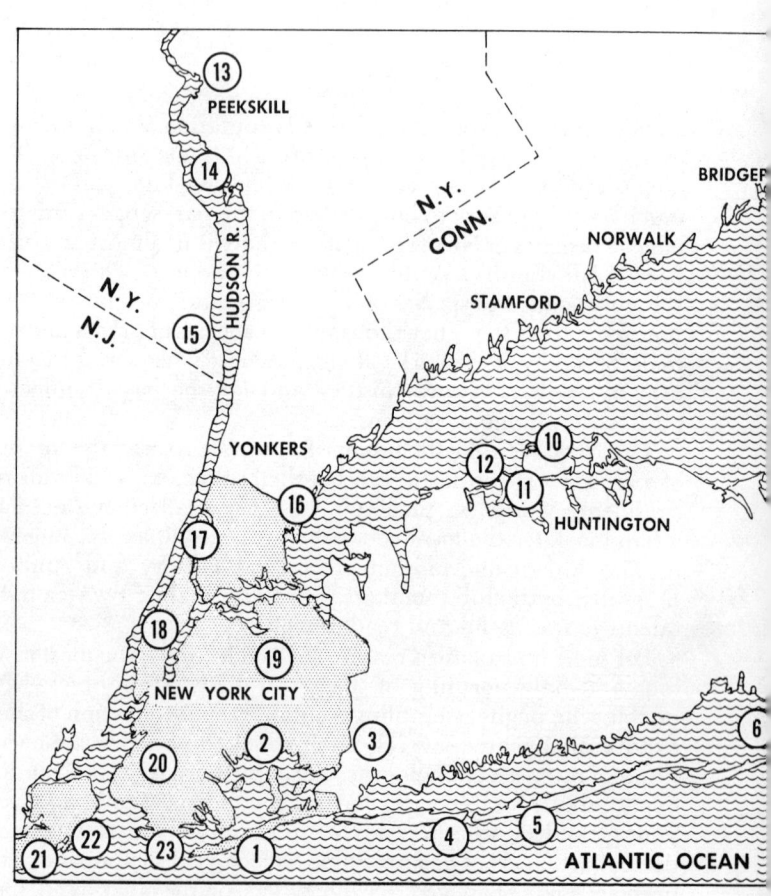

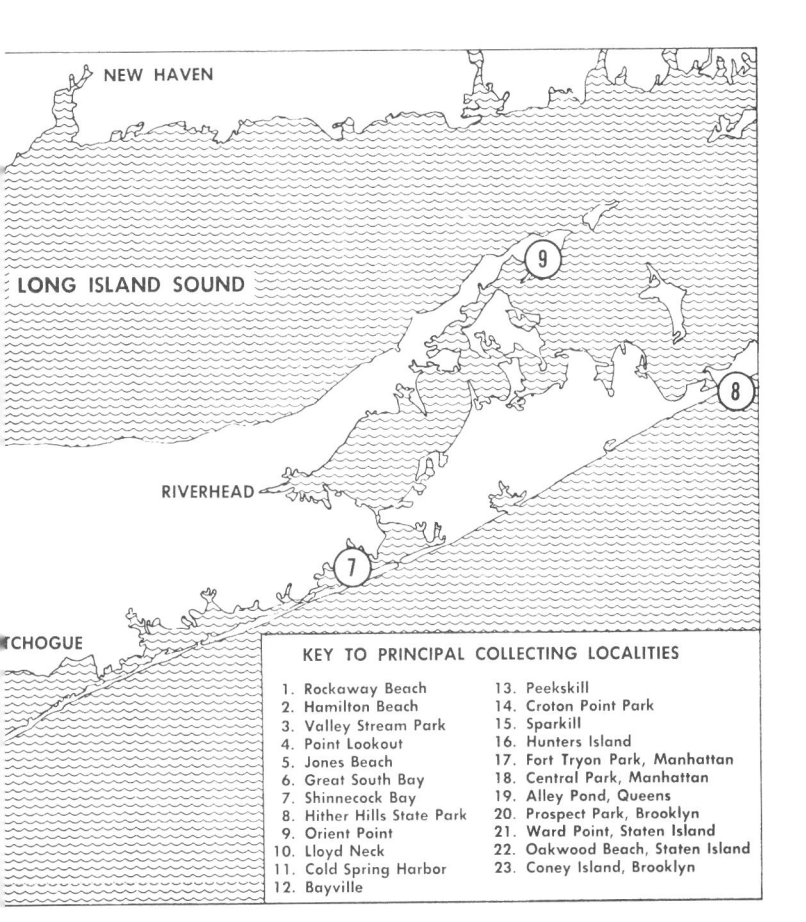

Contents

Introduction   xi

The Land Snails   1
    Triodopsis albolabris, 3 · T. tridentata, 4 · T. denotata, 4 Mesodon thyroidus, 5 · Ventridens ligera, 6 · Stenotrema fraternum, 7 · S. hirsutum, 7 · Haplotrema concavum, 8 · Anguispira alternata, 9 · Succinea ovalis, 10 · S. avara, 10 · Cepaea nemoralis, 11 · Zonitoides arboreus, 12 · Z. nitidus, 13 · Retinella electrina, 14 · Oxychilus cellarius, 14 · Discus cronkhitei catskillensis, 15 · D. rotundatus, 16 · Helicodiscus parallelus, 17 Cionella lubrica, 18 · Vallonia pulchella, 19 · Carychium exiguum, 20 · Pupillidae, 21 · Euconulus fulvus, 22 · Strobilops labyrinthica, 23 · Hawaiia minuscula, 24 · Punctum minutissimum, 24

Mollusks of Fresh Water 25

> Lymnaea palustris elodes, 26 · L. columella, 26 · L. auricularia, 26 · L. humilis, 27 · Helisoma trivolvis, 28 · H. anceps, 29 Planorbula jenksii, 29 · Gyraulus parvus, 30 · Promenetus hudsonicus, 30 · Physa heterostropha, 31 · Aplexa hypnorum, 32 Ferrissia fusca, 32 · Goniobasis virginica, 34 · Viviparus malleatus, 34 · V. contectoides, 35 · Campeloma decisum, 36 · Bulimus tentaculatus, 36 · Amnicola limosa, 37 · Valvata tricarinata, 37 · Anodonta cataracta, 38 · Elliptio complanatus, 39 Margaritana margaritifera, 40 · Sphaeriidae, 41

Marine Gastropods (Univalves) 43

> Acmaea testudinalis, 44 · Epitonium rupicola, 45 · Epitonium *species*, 46 · Polinices heros, 46 · P. duplicatus, 48 · P. triseriatus, 48 · Crepidula fornicata, 49 · C. plana, 50 · C. convexa, 50 Crucibulum striatum, 51 · Littorina littorea, 52 · L. irrorata, 52 L. saxatilis, 53 · L. obtusata, 53 · Lacuna vincta, 54 · Urosalpinx cinerea, 54 · Eupleura caudata, 55 · Thais lapillus, 56 Anachis avara, 56 · Mitrella lunata, 57 · Nassarius obsoletus, 58 N. trivittatus, 59 · N. vibex, 60 · Buccinum undatum, 60 · Colus islandicus, 61 · Busycon canaliculatum, 62 · B. carica, 63 Haminoea solitaria, 64 · Melampus bidentatus, 64 · Ovatella myosotis, 65 · Bittium alternatum, 66 · Cerithiopsis greenii, 66 Seila adamsi, 66 · Triphora perversa nigrocincta, 67 · Odostomia *species*, 67 · Turbonilla *species*, 68 · Hydrobia minuta, 68 Mangelia *species*, 69 · Retusa canaliculata, 69 · Cylichna oryza, 70 · Acteon punctostriatus, 70 · Spirula spirula, 70

Marine Pelecypods (Bivalves) 71

> Solemya velum, 73 · Nucula proxima, 73 · Yoldia limatula, 74 Anadara pexata, 74 · A. transversa, 75 · Noetia ponderosa, 76 Crassostrea virginica, 76 · Pecten irradians, 77 · P. magellanicus, 78 · Anomia simplex, 79 · Mytilus edulis, 80 · Modiolus modiolus, 82 · Brachidontes demissus plicatulus, 82 · Congeria leucophaeata, 83 · Pandora gouldiana, 83 · Periploma leanum, 84 · Lyonsia hyalina, 84 · Arctica islandica, 85 · Astarte castanea, 86 · Venericardia borealis, 87 · Divaricella quadrisulcata, 87 · Laevicardium mortoni, 88 · Venus mercenaria, 88 · Pitar morrhuana, 89 · Gemma gemma, 90 · Tellina agilis, 92 · Macoma balthica, 92 · Donax fossor, 93 · Tagelus plebeius, 94 Siliqua costata, 94 · Ensis directus, 94 · Spisula solidissima, 96

Mulinia lateralis, 96 • Mesodesma arctatum, 97 • Hiatella arctica, 98 • Corbula contracta, 98 • Mya arenaria, 99 • Cyrtopleura costata, 100 • Petricola pholadiformis, 100 • Barnea truncata, 101 • Zirfaea crispata, 102 • Teredo *species,* 102 • Chaetopleura apiculata, 103

Collecting Shells in Fish Stores 105

Otala lactea, 106 • Eobania vermiculata, 107 • Helix pomatia, 107 • H. aspersa, 107 • Theba pisana, 108 • Helix aperta, 109

Shell Lists of Special Areas 111

Coney Island, 112 • Rockaway Beach (*also* Jones Beach, Point Lookout, Fire Island, Long Beach, etc.), 112 • Hamilton Beach (*also* Roxbury, Broad Channel, Riis Park, Canarsie, Bergen Beach, etc.), 114 • Ward Point, 115 • Bayville (*also* Wading River, Sunken Meadow, Huntington, Orient Point, etc.), 116 Sparkill, 117

Glossary 119

Diagram of gastropod, 121 • Diagram of pelecypod, 123

Taxonomy of the Mollusks 125

Supplement of Additional Species 137

Bibliography 145

Index 151

Introduction

In recent years there has been a greatly revived interest in that genteel diversion of earlier days, the accumulation of sea shells in private cabinets. This interest has been reflected in a number of handbooks on shells that have appeared and enjoyed popular acceptance. However, the authors offer no apology for the publication of still another shell book in a market that is rapidly becoming crowded, for our work is designed to fill a need that the other books only imperfectly regard as their concern.

The works of Percy Morris, Maxwell Smith, Tucker Abbott and others are mainly concerned with marine shells and lend a heavy preponderance to the beautiful species that are found only on the subtropical beaches of Florida and southern California. For the vacationist and visitor to these places they are ideal. But in spite of the greatly increased flow of northerners to Miami and points south, it still remains true that the majority of our citizens north of the Mason-Dixon line must content themselves with summer trips to less romantic local beaches. Should any of these stay-at-homes develop a healthy interest in shells, and in pursuit of this interest invest in one of the current handbooks, then it must follow that they can procure the huge majority of the enticing Florida shells only through purchase, exchange or by dunning more fortunate friends and relatives to bring back shells for them. At any rate, they will have to forego the health-bringing aspects of this (all) year-'round hobby, the venturing into the outdoors to hunt down elusive mollusks in their native habitat.

We have compiled our work with these people in mind. Our book aims to bring such people out into the open, so that they may search for shells in their own back yards, so to speak. For this reason we do not confine ourselves to the beaches, but lure them inland to wooded areas and ponds and streams to collect, as well, the mollusks that are found on land and in fresh water. All the shells listed herein can be taken in a radius of about fifty miles from Times Square, with the special inclusion of all of Long Island to Montauk Point. However, most of the shells described and illustrated can be found within or very near the confines of the great city itself, on the beaches, in the bays, the ponds, streams, and the unimproved land areas on the outskirts. For in such areas reside the local members of the phylum Mollusca, and a little practice will soon enable one to recognize good snail territory at a glance.

WHERE AND HOW TO COLLECT

It is useful to know the areas that mollusks are fond of inhabiting, but even in such areas you must know exactly where to look. The remarks that follow will give broad general hints as to specific habitats, but only actual experience will make you a real "snail hunter."

The main point to be kept in mind is that our land mollusks are for the most part shy, nocturnal creatures. During the day they hide away from the sunlight under leaves, boards, stones, or old logs. Hence, to find them you must learn to grub for them under such objects. However, not any old board will do. There should be a certain amount of moisture and an area of relatively undisturbed grassy or forested ground. You can look in such places with a flashlight at night, when collecting is really good; or you could wait for a dank, drizzly day, especially if the night before has enjoyed some heavy thundershowers—for on such a day our snail friends throw caution to the winds and wander abroad to search for tasty snail morsels.

To secure a catch of the minutae (for many of our land snails are of this nature), the best trick is to fill your kit with the moist rotted leaves and light humus that such creatures haunt, dry the "take" at home and then, on some stormy, blizzardy winter evenings, go snail collecting right in your own workroom under a brilliant electric light, with the aid of an ordinary hand lens. This trick serves as well for the collection of minute marine shells.

The mollusks that live in fresh water are easier to obtain. They are out in the open in clear water, where you can see them, and they can be caught with groping fingers or scooped up in a large kitchen strainer tied to a length of stick. Some species, however, hide under lily pads and other water plants, and these must be turned and examined; the tinier animals like to lose themselves in the branching fronds of aquatic plants. Here the technique indicated above is again brought into play. Pull up the plants, take them home to dry, and search at your leisure. The fresh water Sphaeriidae (fingernail clams) live in mud and can be gotten in numbers only by scooping and sifting. The large river clams, which live in lakes as well and are romantically called "naiades"* by zoologists, can actually be seen in clear water, nearly buried in the bottom mud, but with the upper end projecting just enough to enable a diver to get a firm grip and pull them up. Where the water is muddy with suspended silt, you must wade in and feel about blindly with bare hands and feet. Of course the dead shells of all these mollusks can easily be picked up on the banks of streams and ponds and in dried river beds.

Hardly anything needs to be said about marine shell collecting. Just walk on the beach, head and eyes down, and pick up what you see. But even here it is well to pay heed to at least two important points. It is obvious that collecting at low tide is better than at high and that the really minus tides are best of all, for then many species can be obtained alive which otherwise are found only as bleached, worn specimens. Here again

---

* After *Naias*, an old name for the Unionidae.

the technique calls for turning over stones, peering into the crevices of rock and wood pilings, digging into sand and closely examining seaweed and bits of water-logged wood.

The other point to be remembered is that many species are seasonal and some, that either become scarce or vanish completely in certain months, appear in bewildering abundance in others. Hence, there is a need for frequent, persistent collecting tours. Even within a "season," collecting will be affected by the weather. Many books recommend beach visits on days that follow strong on-shore winds. We have found that the collecting of minute snails was particularly good on the day following a strong north winter wind which, by beating down the waves, compels them to deposit gently on the shore the tiny, fragile specimens which otherwise they would smash and fragmentize in their exuberance.

However, there are other species that live just beyond extreme low water and for various reasons are never found on shore. If wading in with a simple strainer or clam hook will not serve in such instances, it then becomes a complicated matter that is better left to experts. It begins to involve iron dredges and chain lengths and lines of half-inch manila rope and lost equipment, or at a more advanced stage, when the mollusks that inhabit the truly bathyspheric depths of the ocean are sought, elaborate equipment that only a Beebe or a Picard knows how to handle. If these remarks have not discouraged the persistence of some collectors, the authors refer them to almost any handbook on marine collecting, where a rich array of dredges and equipment is described and recommended, enough to satisfy the most ambitious student.

NAMES AND NOMENCLATURE

"Common names may change from place to place, scientific names from time to time." In all cases, we have followed sound professional practice by giving the scientific binomial of the animal and the author who first described the species and thus originated the name. Even though we are well aware that such names are generally regarded as "hard," we offer as defense the plea that there simply are no others. Birds and flowers are noticeable enough to have aroused the observational powers of so many people that a usable system of popular names has developed.

But the harmless, shy, retiring tribe of mollusks has never, with some outstanding exceptions, enjoyed the close attention of the layman.

Hence, the few popular names in use, such as oyster, clam, skimmer (a clam in New York but a bird in New England!), periwinkle, conch, coquina, whelk, steamer, hardshell, cherrystone, blue point*(the last three all referring to the same species), are completely inadequate to designate the bewilderingly numerous species of mollusks, second in numbers only to the vast army of insects. Moreover, even these popular names do not always refer to the same animal: thus a conch in New York is a species of *Busycon,* whereas in Florida it is a *Strombus;* a whelk in England is a *Buccinum,* here a *Busycon;* periwinkles are either *Littorina littorea* or *L. irrorata,* or even a flower! And so on. Nevertheless, we have permitted ourselves to suggest "popular" names in our discussions, but we offer these apologetically, knowing full well that few people, even among the initiated, will know exactly what is referred to. Dr. Abbott in his *American Seashells* (see bibliography) has equipped our marine species with a more or less complete nomenclature of popular names, but in the few years the work has been with us, not many of these names have been generally accepted.

We must not leave the subject of names without referring to the seemingly deplorable habit of professional malacologists who change the names of certain shells that have been so laboriously mastered by the beginners. Luckily, comparatively few names are involved. As to the reason for the changes, suffice it to say that the whole subject is so vast that a huge body of legal literature has grown up, as well as an international commission of judges who meet regularly to decide knotty questions of nomenclature. The Keep-Bailey book, *West Coast Shells,* has a short but enlightening chapter on the subject.

PRONUNCIATION

How does one say these more or less formidable scientific names? We freely admit the difficulty, but we feel we should also point out that in our polynational population there are many whose family names are every bit as difficult. Nevertheless, we learn to manage. So it is with the names in zoology. There should be some comfort in the fact that, strictly speak-

---

* The term "blue point" commonly refers to the edible oyster.

ing, there is no single "correct" pronunciation, provided no essential part of the word is omitted; one pronunciation is as valid as another. As to memorizing scientific names, this is done as in all cases by constant and, we dare hope, pleasurable use.

STORING AND ARRANGING THE COLLECTION

Shells are much more easily stored in collections than are flowers or insects, but there is nevertheless a little work involved. A shell which still retains the original inhabitant is best plunged into boiling water for a minute or two, and then the animal may be extracted with a dissecting needle or a small hook bent out of a pin or needle. If the entire mass does not come out in one piece, and persistent fishing fails to catch the left-overs, immersion of the specimen in alcohol for a day or two will frequently be sufficient to keep the shell from becoming offensive. There are several finer points involved in storing shells such as cataloguing, labeling, construction of cabinets, etc. Several excellent papers treat such subjects in great and authoritative detail, among the best being the paper of Dr. William Dall, first published in 1892 and still available for reference in libraries. More recently this topic has been discussed by Dr. Abbott in *American Seashells*.

SCOPE AND PURPOSE

Assiduous pursuit of the delightful hobby of shell collecting in and around New York City for twenty years has given the senior author a pretty good idea of what this area has to offer in the way of mollusks, and we drew heavily upon this accumulated experience to select the species for description in this book. However, we freely consulted the works and faunal lists of such outstanding workers in this area as Temple Prime, Sanderson Smith, Totten, Pilsbry and others.

From this mass of knowledge we have picked for discussion the animals that the collector is most likely to meet, omitting the species that are of doubtful or rare occurence in the region. In the case of the minute shells, where identification is not always easy and long exhaustive descriptions are frequently needed, we have contented ourselves with describing the genus alone and illustrating some of the more common species, so that the user of this book will at least know that he is dealing with a pupillid or a pyramidellid and so on. These minute shells, especially the latter, have been known to give even experts much trouble. However we do not mean to let this act as a deterrent to the ambitious collector; there are excellent texts that treat with these tiny mollusks and the field for original work is wide open.

Elsewhere we have described this Lilliputian molluscan world as containing the most interesting of our snails. For nowhere else does one find such a rich diversity of shape, design and sculpture and such large gaps in our knowledge of the life history of practically all of them, gaps that even the veriest tyro can do much to fill. Hence it is hoped that the sketchy treatment we accord these snails will not serve to deter anyone from their study.

ARRANGEMENT

We owe our readers a few remarks regarding the arrangement we have followed. It would have been an easy matter to adhere to a scientific taxonomic system, even though not all authorities are agreed on any single one. However, after much thought, we have decided to deviate from such an arrangement in a number of important particulars.

In all, we are dealing with about 140 species, a very small number when compared to the vast host of living mollusks, and hence departure from the system will not be too greatly noticed nor lead to any confusion. Our greatest departure is in the case of the land shells, where we guide the beginner from large shells, to smaller, to smallest. Since we devote one section to fresh water shells, we had to remove the fresh water bivalves from their natural position among the other bivalves and place them with the fresh water gastropods. *Petricola*, a marine borer, was also removed from his natural placement and is included with the other borers, *Barnea*, *Zirfaea* and *Teredo*.

In other respects our arrangement of the marine bivalves is in accordance with the generally accepted system in Johnson's check-list. The same can as a whole be said of the marine univalves, except that we removed all the tiny species from their normal position and put them in a sub-section by themselves. In addition we made a few other minor shifts. We hope that beginners will find this arrangement more useful when they are collecting in the field.

For the scientifically-oriented, we have appended a cross-referenced listing by class, order, superfamily, etc., of all the species included in this book. (See pages 125–135.)

With these few remarks we submit this work for your approval, taking comfort mainly in the fact that it will serve to call attention to a much neglected portion of our regional fauna. And for those who find peace of mind, rest, and quiet pleasure in the closer contemplation of the riches of Nature, there should be no overlooked secrets, no unobserved mysteries. For the omissions and errors that undoubtedly have escaped detection, we beg your indulgence, for we intend this work merely as a sesame to the door that, unlocked, will reveal to you stronger, more complete guides to lead you further on.

ACKNOWLEDGMENTS

Even so modest an undertaking as the present one cannot be completed without the sympathetic interest and support of many well-wishers. Chief among these is Dr. Walter H. Jacobs of the New York Shell Club, without whose urging and enthusiasm for shells this work would never even have been begun. Dr. William J. Clench of Harvard, one of the world's leading authorities on mollusks has, with his customary kindness, read the manuscript and offered helpful suggestions. Mr. Frederick Weir of The American Museum of Natural History went to great trouble to provide us with definitive specimens from the rich museum collection and in other ways did much to make the work on the book a pleasure. Dr. Henny E. Coomans and Mr. William E. Old Jr., also of the Museum staff, generously made a final check of the proofs. To these and to all others who helped us with words of encouragement and advice, particularly enthusiastic members of the New York Shell Club, our warmest thanks.

Shells
from Cape Cod to Cape May
WITH SPECIAL REFERENCE TO THE NEW YORK CITY AREA

Collecting land snails is the most difficult and, some think, the least rewarding phase of one's shelling activities. The snails are generally not easy to find; a great deal of stooping or kneeling and log turning and grubbing and sifting is involved. When finally the specimens are procured, it must by and large (except by some) be admitted that they are not very handsome, not at all like the marine shells of our immediate area or the land shells of tropical lands. There are very few large-sized species, the majority being small to tiny, which, of course, fails to add to the lure of land shell collecting.

In addition, hunting the snail can be dangerous—not that the cornered snail will turn fiercely and prepare to sell its life dearly, but rather through attendant phenomena of which our mollusk wittingly or unwittingly is the beneficiary. If, in the area we are searching, the presence of poisonous snakes is known, all the precautions urged upon us by experts when we venture into the forests must be borne in mind, especially since in our hunt we are forced to explore those sheltered, hidden nooks that snakes as well as mollusks generally frequent.

Although it must be admitted that this danger in our well-citified region is excessively rare, yet there is another that is practically omnipresent. In our eagerness, we tend to pay little attention to the shrubbery we are compelled to examine or push aside while hunting, and so frequently come into painful, itchy contact with nettles or, worse, poison ivy. It helps to look about carefully before the search begins and to note the presence of these noxious weeds, but many find it difficult to keep the awareness alive once the fever of the chase begins to burn. Yet the taking of all practical precautions is very advisable.

However, all these difficulties are bound to add zest to the land snail search, especially since it lures us to dank and shady forest glades near singing brooks, where only stray beams of sunshine ever enter and the rich, indescribable odor of fertile humus surrounds us. In such a spot we gather, in addition to the few successful specimens, a spirit of rest and coolness and calm that serves to refresh us even after we have returned to the noisy haunts of man. And every time we see in our cabinet the snail we have collected, we are transported back in mind to the bewitching surroundings where we found it—truly a "souvenir."

TRIODOPSIS ALBOLABRIS Say
White-lipped Forest Snail

SIZE: Diameter just over 1 inch; height ½ inch.

DESCRIPTION: The shell is dully shining or matte, cream buff or chamois color, often darker. It is usually so thin that the various organs can be seen within. The spire is depressed, and a thick, white reflected lip completely covers the umbilicus. In fresh specimens the lip is sometimes a faint purplish orange color which soon fades.

The white-lipped forest snail is the largest and one of the most common of our native land shells. It is found in partially cleared forests in the country areas near New York and lives sheltered in moist mold under decaying logs and rotten stumps. Frequently it can be found on empty lots under discarded corrugated pasteboards. At times it can be found about stone walls and rocks in open fields. The characteristic lip is not developed until the snail is mature, hence young specimens lacking this structure often seem to be a distinct species. When found dead, the shell is white, opaque and chalky, owing to the loss of the living tissue of the shelly material.

One of the best known American conchologists, Dr. Augustus A. Gould, wrote engagingly in 1841 about *T. albolabris:* "This is our largest snail, and, although so simple in its structure and coloring, is a pleasing shell. Its delicately striated surface, and broad, white lip, cannot fail to gain admiration . . . The economy of these animals may be briefly stated as follows. They subsist upon decaying leaves and vegetable fibre, under which they usually shelter themselves. In moist weather, and after showers, they issue from their retreats, and crawl over the leaves or up the trunks of trees, until driven back by a change of weather. In early spring they are often seen collected in groups on the sunny side of rocks. In June they deposit their eggs to the number of thirty to eighty in the light mould by the side of rocks and logs. These are white, opaque, and elastic; and in about twenty to thirty days the young animal issues from them with a shell consisting of one whorl and a half. In October they cease to feed, and select a place under some log or stone where they may be sheltered for the winter, and there they fix themselves, with the mouth upwards. This they close by secreting a thin, transparent membrane, and as the weather becomes cold, they grow torpid, and remain in that state until the warmth of spring excites them to break down the barrier, and enter upon a new campaign of duty and pleasure."

The variety TRIODOPSIS ALBOLABRIS TRAVERSENSIS Leach is easily separated from *T. albolabris* by its noticeably smaller size, but according to Dr. Pilsbry of the Philadelphia Academy of Natural Sciences, it differs in certain anatomical particulars as well. We have found it in large numbers on the walls and under overhangs in the well-known Annsville Cut near Peekskill, as well as in the wooded country nearby. On the walls of the Cut, a large roadway blasted through the heart of a mountain to accommodate the short road that connects the Bear Mountain Highway and Route 9, the snails are found in the daytime near spots kept moist by seeping ground water. This snail has also been collected at Staten Island and on Long Island at Sea Cliff, Huntington and Orient Point.

Three-toothed Forest Snail
TRIODOPSIS TRIDENTATA Say

SIZE: ½ inch in diameter, ¼ inch high.
DESCRIPTION: The shell is very depressed and has a large, open umbilicus. It is cinnamon in color and slightly glossy. The lip is narrow, with three teeth, two of which grow from the sides of the outer lip, the third and largest from the parietal wall.

The three-toothed forest snail is much shyer than the white-lipped snail and hence a good deal harder to find. We have found it near Peekskill after wearily winnowing through large masses of moist and decaying leaves piled up in deep windrows against a stone fence. At times it is found under loose rocks and under rotting logs. It is found in many localities but is not common anywhere in our area.

Hirsute Forest Snail
TRIODOPSIS DENOTATA Férussac

SIZE: ⅞ inch in diameter, about ½ inch high.
DESCRIPTION: The shell is depressed, snuff brown and quite dark in color; in life covered by a dark brown, fuzzy

periostracum. The umbilicus is usually covered by the extension of the rather wide, white lip. The aperture is contracted by two teeth projecting from the inner margin of the outer lip and by a prominent curved tooth on the parietal wall.

This snail, which is most easily recognized by its characteristic periostracum, seems to be the rarest of the larger land shells in this area. We have found several specimens at Lake Peekskill in Westchester County and at Palisades, New Jersey, mostly by the light of a flashlight on humid nights. It is found in damp places under rotting leaves and logs.

In older books this snail is called *Helix* or *Polygyra palliata* Say, but Dr. Pilsbry showed that this name had already been used in 1807 for an Alpine snail. Hence the next valid name, that of G. P. Deshayes, proposed in 1830, had to be substituted. However, this was in turn superseded by Férussac's designation.

Common White-lipped Snail
MESODON THYROIDUS Say

SIZE: ¾ to ⅞ inch in diameter, about ½ inch high.

DESCRIPTION: The shell is moderately depressed, rather thin. The ivory yellow surface is somewhat glossy, with fine oblique lines and very weak spiral lines. The lip is white, moderately wide, and does not cover the umbilicus; there is sometimes a short, weak tooth on the parietal wall.

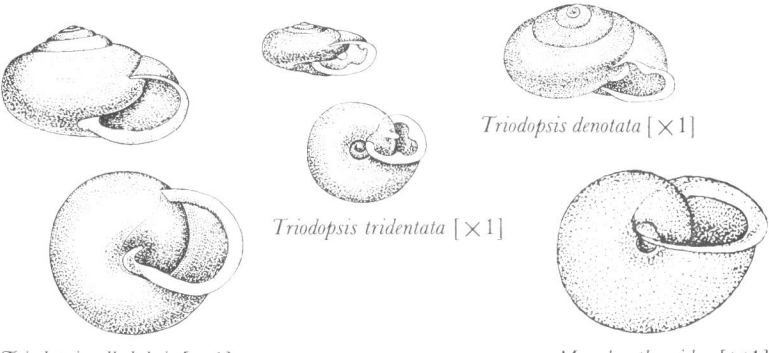

Triodopsis denotata [×1]

Triodopsis tridentata [×1]

Triodopsis albolabris [×1]

Mesodon thyroidus [×1]

This snail, closely resembling *Triodopsis albolabris,* is readily distinguished by the uncovered umbilicus, the adjacent area of the lip merely providing a sort of slanting shield under which the opening can easily be seen. The specimens in our area are quite small, but we have collected larger animals in neighboring Connecticut, and in other parts of the country the shell is almost as big as *T. albolabris*. *M. thyroidus* is frequently taken in the parks of Queens and Staten Island, in moist situations under leaves and rotting logs.

We once took home a number of immature specimens from near Alley Pond in Douglaston and kept them alive in a small vivarium, feeding them fresh crisp lettuce. They took to this food avidly at first, but in time their appetite seemed to falter and, most distressing of all, the white lip, the sign of their maturity, failed to appear. On another excursion, we brought them a bagful of moist, richly rotted forest leaves. As much as such an action can be ascribed to snails, they fell voraciously upon the "natural" food and, coincidence or not, in a very few days a faint lip appeared, and in a week or so the snail shells had reached triumphant maturity.

VENTRIDENS LIGERA Say *Light Belly-tooth Snail*

SIZE: About ½ inch in diameter, ¼ inch high.

DESCRIPTION: The polished yellowish shell is rather elevated, with narrow whorls on the spire and a large, well-rounded body whorl. The base is gleaming and deeply indented about the small umbilicus. There is an opaque buff or light yellow spot behind the basal lip.

This species, which is widely distributed in the northern and central states, from New York to Arkansas and Oklahoma, is recorded in our area only from Staten Island. Here a number of years ago the New York Shell Club on one of its annual excursions found a huge colony at Springville near Travis Road, on and near a plowed but fallow field. Its pleasing shape and bright, shiny color make it one of our handsomer land shells. The local species probably belongs to the form called *stonei* Pilsbry, which is characterized by being thinner and smaller than the typical. Its shape is so distinct that there is no danger of its being mistaken for another species.*

---

* Another *Ventridens* species is described in the Supplement of Additional Species, page 137.

STENOTREMA FRATERNUM Say   *Flat Pill Snail*

SIZE: Nearly ½ inch in diameter.
DESCRIPTION: The shell is depressed, slightly convex, with five or six whorls that diminish very slowly in width from the outer whorl to the apex. The dark russet or chestnut colored periostracum has very tiny fuzzy hair-like projections. The aperture is long and narrow and is constricted by a rather long, white, tooth-like ridge on the parietal wall. The lower lip is never notched. The umbilicus is nearly or altogether covered by the strongly reflected, narrow lip. It is larger and flatter than the next species (*S. hirsutum*) with which it need never be confused.

This snail is not uncommon around Peekskill and on Long Island where, being apparently somewhat bolder than its smaller relative *S. hirsutum*, it lives rather more openly. We once found a round dozen clinging to the trunk of a large tree, the shell difficult to see against the rough bark. It can also be found on forest floors under light leaf mold or clinging to the underside of fallen tree trunks and loose bark.

STENOTREMA HIRSUTUM Say   *Hirsute Pill Snail*

SIZE: Slightly more than ¼ inch in diameter and about ⅜ inch high. DESCRIPTION: The shell is nearly globular, dull brownish in color with about five rounded whorls. There

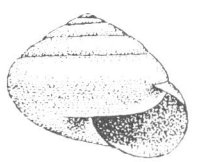
Ventridens ligera [×3]

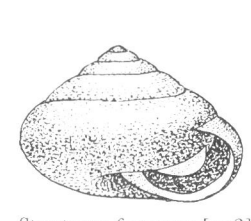
Stenotrema fraternum [×3]

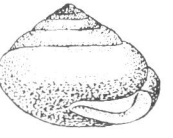

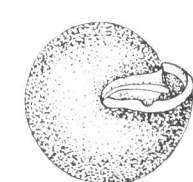
Stenotrema hirsutum [×3]

is no umbilicus, but the umbilical region is deeply indented. The aperture is narrow and much obstructed by a long parietal tooth that almost covers the view of the opening when looked at from the side. The lower lip is raised and has a small notch located a bit nearer the columella than the peristome. Its outstanding characteristic is the hairy quality of its periostracum from which it derives its name. The hairs, which are short but clearly visible under slight magnification, are fragile and easily rubbed off. They are of course never seen on dead and bleached specimens.

This typical forest dweller is not common near the city. It has been taken under moist leaves, especially at the base of boulders and under fallen tree trunks, near Peekskill. In 1891, a colony of this snail was recorded "in a small patch of woods" at what is now 13th Avenue and 74th Street, Bay Ridge, Brooklyn. When we visited this spot recently, we found only concrete, bricks and steel occupying the location. Because of its small size and dark, lusterless color, *S. hirsutum* is not easily seen on the forest floor and is frequently passed up by the beginner as a piece of dried mud.

HAPLOTREMA CONCAVUM Say — *Cannibal Snail*

SIZE: In the New York area from ½ to ⅝ inch in diameter, about ¼ inch high. DESCRIPTION: The depressed, dully shining shell is of a yellow color and has a deep, wide umbilicus that shows all the whorls, including the last one. It has a rounded, toothless aperture and, when mature, has the lip very slightly thickened and barely reflected. A peculiar feature is the slight depression on the top edge of the last whorl not far from the suture, which produces a pronounced dip in the upper lip.

This snail, which Dr. Frank C. Baker properly called the tiger of the molluscan world, subsists entirely on live prey and preferably on other mollusks. It is found occasionally on Long Island and around Peekskill, where we have sometimes taken it in the very act of feeding. The animal is long and slender and is thus able to plunge fiercely into the aperture of a retracted food snail, where it consumes its victim to the last morsel in the topmost whorl. We once found a large *Haplotrema* dining on an immature

Triodopsis and, in its eagerness, it was devouring shell and all. There is reason to believe that it will even dine cannibalistically on others of its own species.

ANGUISPIRA ALTERNATA Say   *Striped Forest Snail*

SIZE: Although it grows to about an inch in diameter, the specimens around New York City are frequently smaller, most of them being a little more than one-half that size and about ¼ inch high. DESCRIPTION: The shell is depressed, with a wide and deep umbilicus that shows all the whorls. The lip is sharp and thin and never develops a reflected edge. It is rather prettily colored with irregular, oblique, brownish dots and dashes on a light buff background, a type of marking that can be characterized as flamelike. Its sculpture consists of closely spaced ribs that appear quite coarse when magnified. Some specimens are more depressed, with a slight ridge at the outer periphery.

This snail prefers to live in colonies, so that once one is taken, it is comparatively easy to obtain a good series. We found it near Alley Pond in Douglaston, Queens, and in practically every wood lot anywhere that has not been too thoroughly improved. It is a native snail, and seems to be quite as successful in maintaining itself in suburban environments as the introduced *Cepaea nemoralis* and the slug *Limax maximus*.

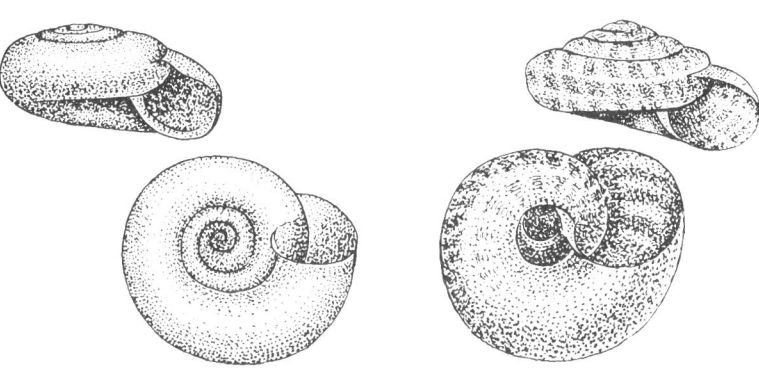

Haplotrema concavum [×2]   *Anguispira alternata* [×2]

A beautiful subspecies of this snail has been reported near the Greenwood Cemetery and elsewhere in Bay Ridge, Brooklyn. It is comparatively smooth, somewhat glossy and of a pale yellow color with deep red-brown flame markings and spots in three distinct spiral series. It is called ANGUISPIRA ALTERNATA FERGUSONI Bland, after an active Brooklyn conchologist who left his collection to Columbia University. It is also found in the New York Bay Cemetery in Jersey City, Ruckman's Ravine in the Palisades and other localities.

SUCCINEA OVALIS Say *Oval Amber Snail*

SIZE: ¾ inch high, a little less than ½ inch wide.
DESCRIPTION: The body whorl is very large, with an aperture that is well over three-fourths the size of the entire shell. The shell is greenish smoky amber in color, and so thin that the spotted mantle of the animal clearly shines through. The lip is thin and quite fragile and the spire very small but distinctly elevated.

The amber snails haunt moist places not far from ponds and streams and in dry weather hide under sod, leaves and stones. When the weather is rainy they climb up plants and trees; on such a day we once collected many clinging to low poison ivy plants in Croton Point State Park. This brought us a very personal experience with one of the hazards of snail hunting in the New York area. In the winter, *Succinea* retires into loose forest mold and closes the aperture with a thin, transparent epiphragm. It is a common shell and, in our area at least, not easily confused with other species.

SUCCINEA AVARA Say* *Lesser Amber Snail*

SIZE: Up to ½ inch in height, about ¼ inch in width, frequently found much smaller. DESCRIPTION: Shell small, thin, fragile, of a deep straw color; aperture about one-half the size of the shell, the spire rounded and the suture distinct; the lip thin, never thickened or reflected.

The smaller amber snail likes to live in such moist situations that for a time it was not clear whether this was an aquatic or

* See List of Changes in Nomenclature, page iii.

land snail. It is easily distinguished from the previous species by its proportionately smaller aperture, the spire being fully as large as the aperture, instead of being merely a fraction thereof. It is lighter in color and without the smoky tinge of *S. ovalis*. Since it prefers to live on muddy banks of streams, rivers, and ponds, it is usually found covered with a liberal coating of dirt.

The identification of the smaller species of *Succinea* is difficult when consideration is given only to the shells. In very recent years, several workers have begun a thorough study of the soft parts of the animals, and it is likely that several drastic revisions will have to be made in the nomenclature. From unpublished sources, for example, we learn that many of the specimens identified as *S. avara* are actually another species entirely. However, until some degree of certainty is attained, it is better to abide by the old concepts.

CEPAEA NEMORALIS Linné   *English Garden Snail*

SIZE: About ¾ inch in diameter, ½ inch high.
DESCRIPTION: The shell is depressed globular, about the size and shape of *Mesodon thyroidus*. However, it is glossy and much more vividly colored. The ground color may be bright lemon yellow or a light reddish brown, being either a solid color or decorated with from one to five, wide or narrow, dark brown bands. The lip is always brown. There is no umbilicus.

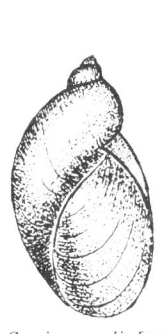

Succinea ovalis [×2]

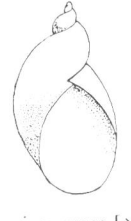

Succinea avara [×2]

Cepaea nemoralis [×2]

The English garden snail is the last of our comfortably large land snails. A really handsome shell, it is an immigrant from Europe where it has a reputation as a garden pest. But in our harsher climate, it has earned only occasional blame from hothouse gardeners. The animal is very gregarious—large, populous colonies have turned up in recent years at Douglaston, Malverne, Flushing, and Rockaway Beach. It seems to appear wherever ornamental shrubs have been imported or introduced from nurseries that deal in such shrubs. Near our home in Rockaway, there were at one time five flourishing colonies on empty lots, each one so isolated by the intervening streets, that they developed distinct group characteristics. However, the building boom caught up with the neighborhood and now the only *Cepaeas* that are to be found in Rockaway are in our own back yard where we had hurriedly transferred a few fugitives from the devouring bulldozers.

ZONITOIDES ARBOREUS Say   *Tree Zonite Snail*

SIZE: Usually less than ¼ inch in diameter.
DESCRIPTION: The shell is small, quite depressed, moderately glossy; the color some shade of dark yellowish amber. It has a small umbilicus and usually five rather narrow whorls. The lip is thin and unreflected.

This very common snail is found just about anywhere that conditions are favorable to molluscan life, even under bits of wood, cardboard and other debris in unimproved city lots. It is also quite common in all forests under loose bark of fallen trees and under the leaves and debris of a forest floor. We have it from many lots in Brooklyn, Queens and Long Island, and practically everywhere near Peekskill.

Zonitoides arboreus enjoys the distinction of being the first native American snail to be described and named in the scientific literature. Thomas Say, who described it in 1816, was born in Philadelphia in 1787 and died in New Harmony, Indiana, in 1835. He was one of the founders of the Philadelphia Academy of Natural Sciences and one of the leading naturalists in our country. He described a large number of our native snails and insects, and it is a tribute to his astuteness and sharpness of vision that so large a proportion of his names are still considered valid, even with the far more thorough and detailed techniques in use today. He was able to look beyond accidental and unimportant characteristics of individual specimens and describe the essentials of the entire species. He is generally known as the father of American conchology.

Zonitoides has been accidentally introduced into many countries of Europe and into the islands of the Caribbean. In Louisiana it is considered a pest by the sugar cane growers, because it gnaws at the delicate root hairs of the growing cane, leaving small holes through which harmful fungi and other parasites may enter to cause serious damage. It is found in hothouses on Long Island, where it causes some financial loss by nibbling at the blooms of orchids and other commercial cut-flowers, thereby making them unmarketable. Because of this unfortunate habit, it has recently come to the attention of plant experts, who are working out methods of eradication or control. It will probably prove to be a stubborn foe, not easily defeated.

ZONITOIDES NITIDUS Müller *Shiny Zonite Snail*

SIZE: Usually just more than ¼ inch in diameter.
DESCRIPTION: Somewhat more elevated and darker than *Z. arboreus*. It is decidedly glossy and has a rounded base with a moderately large umbilicus.

This snail is admittedly not easy to separate from *Z. arboreus* unless the animals are caught alive. *Z. nitidus* haunts very wet situations and is black when the animal is still inside the shell, whereas *Z. arboreus* prefers much dryer places and when alive

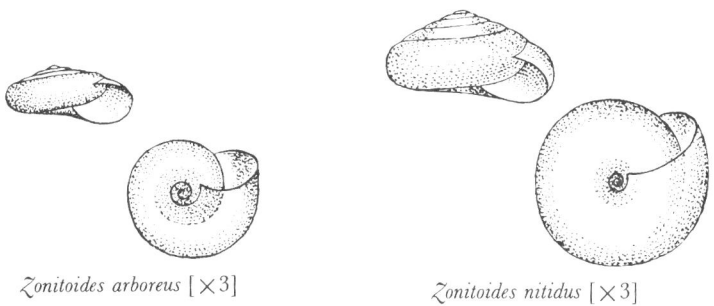

Zonitoides arboreus [×3]   *Zonitoides nitidus* [×3]

appears to be of a light brownish horn color. In addition, *Z. nitidus* is larger, has a somewhat higher spire, a larger umbilicus and a more rounded base. Once the two species are compared while they are still alive, they are never again confused.

We found *Z. nitidus* to be quite common at Alley Pond in Queens under leaves and rubbish at the very edge of the water. Apparently it enjoys being completely immersed at times. It also is found in many northern European countries.

RETINELLA ELECTRINA Gould   *Incised Zonite Snail*

SIZE: Less than ¼ inch in diameter.
DESCRIPTION: The shell is about the size and shape of *Zonitoides arboreus*, with the last whorl widening rapidly so that it is twice as wide as the penultimate whorl. The surface is very glossy, the color very light, just barely tinged with yellowish amber. Under low magnification, there can be seen a number of distantly placed, distinct, impressed radial grooves that run parallel to the growth lines. The presence of these grooves serves to distinguish the shells of the genus *Retinella*.

Since this shell is found in the same habitats as *Zonitoides arboreus*, frequently being associated with it, the two are sometimes confused. However, the high gloss, the light color, the relatively wide body whorl and the characteristic radial grooves readily set *Retinella* apart.

RETINELLA INDENTATA Say is much like the above but has so minute an umbilicus that it appears to be without one entirely. Hence a quick look at the basal area will serve to distinguish the two.

We found both Retinellas, always in small numbers, in almost every bundle of leaf siftings from Peekskill and Long Island we examined.

OXYCHILUS CELLARIUS Müller   *Common Cellar Snail*

SIZE: Just over ¼ inch in diameter, about ⅛ inch high.
DESCRIPTION: The shell is flattened, quite glossy, imperfectly transparent, faintly yellowish in color and more or less distinctly whitish about the small umbilicus.

A rather handsome immigrant from Europe, this snail is strongly addicted to living in the very city itself. Its name is not accidental, for it is reported as inhabiting dank spots in old basements where it is not above hiding near barrels of pickles, wine, potatoes, and other stored foods. It is common in Staten Island, near the Voorlezer's House in Richmond Town. We found it on a lot on MacDonald Avenue directly opposite the Greenwood Cemetery. Fine, large specimens live under loose boulders near the Annsville Cut, described on page 4.

OXYCHILUS DRAPARNALDI Beck, which is found in the open in many situations favored by *O. cellarius,* is not easily distinguished from the latter. It is larger (¾ inch), darker colored and with a wider umbilicus. We have found shells that answer this description at the base of the hill in Fort Tryon Park in beds of pachysandra just above Riverside Drive.*

OXYCHILUS ALLIARIUS Miller is the third of this group of European imports. It is the smallest of the three but differs markedly in possessing a strong garlicky odor, whence its Latin name. It has been reported from Brooklyn.

Common Disk Snail
DISCUS CRONKHITEI CATSKILLENSIS Pilsbry

SIZE: About ¼ inch in diameter.

DESCRIPTION: The shell is greatly depressed, dark brown or coppery, and quite dull in sheen. The umbilicus is very wide, showing the lower portion of each of the whorls to the very apex. Under a magnifying glass, the entire sur-

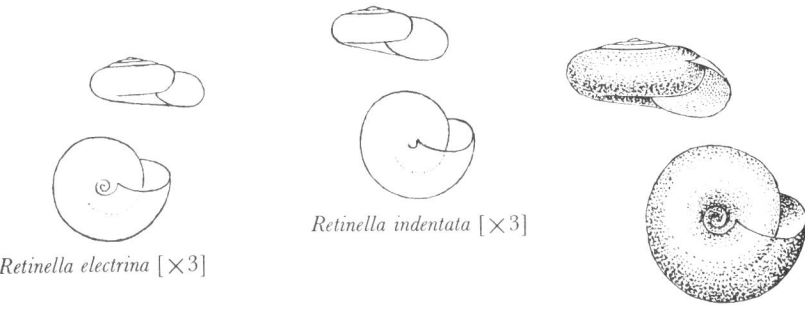

Retinella electrina [×3]

Retinella indentata [×3]

Oxychilus cellarius [×3]

* Two related species, *Mesomphix inornatus* and *M. cupreus,* are described in the Supplement of Additional Species, page 137.

face is seen to be covered with regularly placed, slightly curving axial ribs. The lip is thin and unreflected.

This species is about the same size as *Zonitoides arboreus* but lives in slightly moister habitats. It is easily distinguished by the wide, open umbilicus, the very dull surface and the striking (if minute) surface sculpture. It is quite common in many moist spots under bits of wood, cardboard and stones. We once collected scores of fine specimens in a moldy horse blanket in a dump near Peekskill. Near Palisades, New Jersey, we once collected some white forms of this snail that are called variety *alba*. In older books it is sometimes called *D. cronkhitei anthonyi* Pilsbry.

Among the innumerable localities where we have found *catskillensis*, it was only in Piermont, Rockland County, that we also discovered the typical DISCUS CRONKHITEI CRONKHITEI Newcomb, for which there are also records from Dutchess County. In this subspecies, the entire last whorl is gently and evenly rounded, whereas *catskillensis* is distinguished by a weak but distinct keel or carina along the outer edge of the last whorl.

DISCUS ROTUNDATUS Müller *European Disk Snail*

SIZE: About ¼ inch in diameter.

DESCRIPTION: The shell is very depressed, with six or seven very narrow whorls. The color is yellowish brown, ornamented with regularly placed dull reddish flamelike markings, somewhat like *Anguispira*. The surface is moderately glossy, sculptured with curving, regular, closely-spaced axial riblets. The umbilicus is very wide, open, showing all the whorls to the spire.

This small snail, which is widely distributed in northern Europe, has very recently been found in Fort Tryon Park near West 193rd Street. In addition it has turned up in Rockaway Beach, Great Neck (on the grounds of the old Grace Estate), and in Union City, New Jersey. The snail has the same size and general appearance of our native *D. cronkhitei catskillensis*, but is easily distinguished by its flatter form, its narrower and more numerous whorls, and chiefly by the presence of the brownish spots along the whorls. It lives in very moist situations (as in the pachysandra bed in Fort Tryon Park), or in drier ones (as in Great Neck) or under decaying garden refuse (as in Rockaway). In Rockaway its original habitat has been built over, but fair numbers can be found in the back yard of one of the authors in Rockaway Beach.

These forms of *Discus*, together with *Anguispira* (page 9), belong to the family Endodontidae. This is a huge family of worldwide distribution, with particularly large numbers of species in New Zealand. The tiny eggs of these snails are protected by a white calcareous cover, whereas most of our other land snail eggs are translucent and protected merely by a thin membrane.

Parallel Mulch Snail
HELICODISCUS PARALLELUS Say

SIZE: About ⅛ inch in diameter.
DESCRIPTION: The shell is flat, discoidal, and has four whorls that are equally coiled above and below so that no true umbilicus is present. Inside the aperture can be seen one to three minute tooth-like protuberances. The surface is sculptured with parallel spiral ridges, easily seen under weak magnification. The color is light greenish-yellow.

A woodland dweller that shuns exposed places, this small snail is easily distinguished from others of the same size by its extremely flattened shape and distinctive color. Its outstanding characteristic, from which it gets its name, is a number of concentric ridges encircling the top of the whorls. This feature alone, which is just barely visible to the naked eye, serves at once to identify it. This shy snail, which is probably blind too, can be found only with considerable grubbing in wet leaves and under decayed wood in shady spots in the forest. We have it from Peekskill, Valley Stream, Springville and elsewhere on Staten Island, and from many localities on the New Jersey shore.

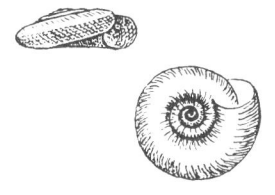

Discus cronkhitei catskillensis [×3]

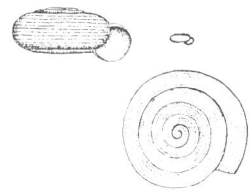

Helicodiscus parallelus [×5]

CIONELLA LUBRICA Müller   *Apple Seed Snail*

SIZE: ¼ inch high, less than ⅛ inch wide.

DESCRIPTION: The shell is cylindrical in shape, smooth, glossy and weakly transparent. The color is pale amber. In living specimens, the animal can be seen through the shell.

The northeastern United States is the only region in the world where practically all the species of larger land snails have shells that are depressed or flat. *Cionella lubrica,* until recently known as *Cochlicopa lubrica,* small as it is, still enjoys the distinction of being the largest elongated land snail in our area. It is a common but quite beautiful little species that appears in Europe as well. At times it can be collected in huge numbers. It is found under bits of board and leaves at the edge of woodlands and even on suburban lots. We once collected a good series from a lot right on the Boulevard in Rockaway. The species is particularly plentiful near the pond at Piermont, in Rockland County, New York. Every lot we examined in Staten Island had a few specimens.

With this species we leave the smaller land snails that can be easily seen with the naked eye. However there are also many that, even when fully developed, are much smaller, some in fact only one or one-and-a-half millimeters in diameter. It would be a pity not to make some mention of these snails, which in many respects are the most interesting in our area. They can best be collected, as we mentioned above, by carrying home the moist leaves and leaf mold in which they live and then searching for them in the siftings. From good localities the haul can be very gratifying. Once, from a small bagful of humus taken in Mt. Olive, Essex County, New Jersey, we collected no less than 132 specimens of these tiny shells comprising fourteen different species.

The question arises as to how one can tell whether the tiny specimen under the magnifying glass is a mature Lilliputian or merely a newly born member of a species that grows much larger. The easiest way is to count the whorls: a mature shell of whatever size possesses at least four or five, whereas the infant shell rarely has more than one-and-a-half or two. Another point to be observed is that the nuclear whorl, the "button" of the spiral, is larger in a juvenile shell in proportion to the succeeding whorls than that of a mature shell.

Although faunal lists of the New York area (see p. 139) give fully as many minute species as larger ones, we shall confine ourselves to a few of the more common forms. The work involved in correctly identifying these specks is frequently quite complicated and calls for considerable experience. It is a wise course for the beginner to enlist the help of experts in the museums and universities, who are usually glad to be of assistance, provided exact locality data accompany the specimens. Chances are very good that even the amateur collector will turn up new geographic records for these frequently overlooked animals and thus contribute his share to the growth of our science.

VALLONIA PULCHELLA Müller   *Handsome Vallonia*

>   SIZE: Minute, about 1/16 inch in diameter.
>   DESCRIPTION: The shell is depressed, glassy white in color, weakly shining, almost transparent. The lip is opaque and well reflected about the roundish aperture. The base is well rounded and the umbilicus comparatively large and deep.

One of the most easily found of the minute snails is this glassy white species that can be taken under moist boards, stones, bricks, shrubbery or flower pots in most city lots and gardens. A board left in a sheltered spot in a back yard in contact with soil and retaining a modicum of moisture on the underside, will in time attract a colony of *Vallonia*. We have found it everywhere

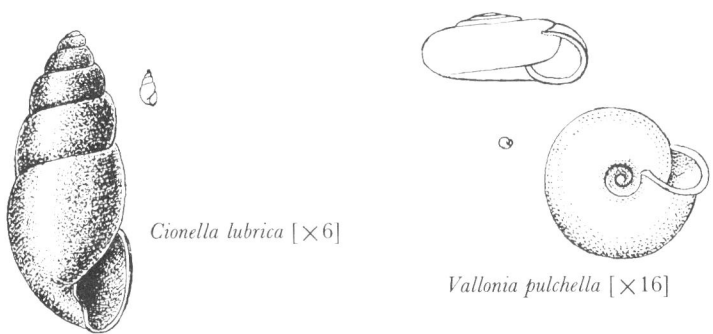

Cionella lubrica [×6]

Vallonia pulchella [×16]

we looked and collected many of them right in our own back yard. Mr. Silas C. Wheat, an old-time collector, who used to live in Sterling Street, Brooklyn, wrote as follows: "In my gardens in the heart of Brooklyn are innumerable *Vallonia pulchella*. I have taken 50 specimens from a space of four inches square. In November I have found them active under half an inch of frozen earth and snow." This snail is widely distributed in the Old World as well as here.

Another species, VALLONIA COSTATA Müller, is very much like *V. pulchella* but its surface is covered by a large number of fine axial ribs or costae (whence the name). Until recently this shell had not been reported as occurring in the New York City area. But a short time ago an active collector of the New York Shell Club found a small colony at the western end of the swamp near the golf course in Van Cortlandt Park, Bronx. This is the first and only record of this snail from New York. Unfortunately the area has been subsequently drained, cleared and improved, so that our incipient colony has probably disappeared.

Staten Island is the type locality of still another species of *Vallonia*, namely VALLONIA EXCENTRICA Sterki. The shell is very much like *V. pulchella* but can be separated from this species, according to Dr. Pilsbry, "by the more oblong contour of the shell and umbilicus, the last whorl widening more towards the aperture . . . and by the smaller and lower spire." However, there is a good deal of controversy regarding the validity of this species. Many experts maintain that this form of *Vallonia* grades gradually into the typical *V. pulchella,* and hence cannot be regarded as a distinct form. Moreover, they state that the characteristic oblong shape can also be found when large series of *V. costata* are examined, and hence seems to be a variation found in the entire genus. Our own experiences do not permit us to decide one way or the other. Presumably the controversy will not be resolved until the internal anatomy of the various species is closely examined.

CARYCHIUM EXIGUUM Say *White Swamp Snail*

SIZE: Minute, about ⅛ inch high, less than one-half as wide. DESCRIPTION: The tiny shell is elevated, narrow, white, shiny or waxy. The lip is reflected. Two small ridges are seen at the aperture, the outer end of lamellae that enter deeply into the body whorl.

This snail, barely visible to the naked eye, shows up beautifully under low magnification. It lives in quite moist situations, never far from streams and ponds in the woods. We have found it on the banks of brooks in Peekskill, in similar situations in Long Island, and at Springville, Staten Island. It lives on very wet dead leaves and must be looked for with some care.

The species CARYCHIUM EXILE H. C. Lea can be separated from the above by the close, distinct and regular striae on the last two whorls. It is often found with *C. exiguum*.

The family Ellobiidae, to which the species of this genus as well as *Melampus* and *Ovatella* belong, is generally regarded as one of the most primitive of the lung-bearing (pulmonate) snails. This primitive condition is shown, among other things, by the fact that the snails, though air breathers and land dwellers, are never found at any great distance from water.

PUPILLIDAE *Pupa Snails*

SIZE: Minute, about ⅛ inch high.

This family comprises a large number of minute cylindrical shells that remind one of diminutive elongated insect pupae. The color is either wax white in the local species of the genus *Gastrocopta* or brownish amber in the genera *Vertigo* and *Columella,* and the aperture is usually much contracted by the presence of sev-

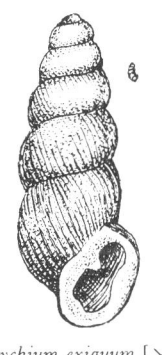
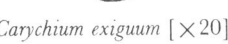
Carychium exiguum [×20]

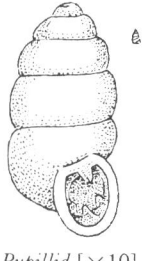

Pupillid [×10]

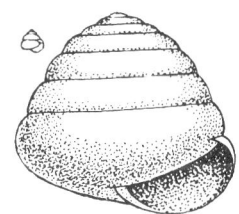
Euconolus fulvus [×8]

eral teeth. The specimens can be collected in large numbers in favorable situations. An adequate description of each of these minute species would make this book too extensive, and we refer the collector to more advanced texts, such as Dr. Pilsbry's monumental monograph on the land mollusca of North America, for a thorough discussion. The authorities in the National Museum in Washington and elsewhere are always ready to identify any of these difficult species for the interested collector.

Vertigo ovata Say, a small, roundish, shiny brown snail is found on dead rushes on the margin of Alley and Lilly Ponds in Queens. *V. gouldii* Binney, less swollen than *V. ovata* but of the same burnished brown color, is a species we once collected on a snowy day in Peekskill where the colony was living under the loosened bark of a dead tree. The various forms of *Gastrocopta* can be found in moist leaf mold from the edges of small forests. *G. contracta* Say, the most common one in our area, is generally covered with dirt when alive and hence not easily found. *Gastrocopta tappaniana* C. B. Adams, a more slender species than *G. contracta*, was recently collected near the bird sanctuary at Springville, Staten Island. *Columella edentula* Draparnaud differs from the other pupillids in lacking the tiny, commonly quite numerous, teeth that adorn the aperture of the others. We found this species in leaf mold from New Jersey.

EUCONULUS FULVUS Müller *Beehive Snail*

SIZE: Minute, ⅛ inch high, slightly more in diameter.
DESCRIPTION: The shell is thin, conic, shiny, light tawny, shaped like a low, old fashioned bee hive. The umbilicus is tiny or closed, the lip sharp.

We have found this shell not infrequently in leaf siftings that also contain other pupillids. Near Peekskill we found a few under the loosened bark of a dead tree. It differs from EUCONULUS CHERSINUS Say, with which it is sometimes confused, by the smaller number of whorls (5 instead of 6 or 6½) which, in addition, are wider. Moreover the body whorl of *E. fulvus* is angular at the periphery, whereas in *E. chersinus* it is well rounded.

Although many old lists give *E. chersinus* as a New York shell, Pilsbry states that "the two species (i.e. *fulvus* and *chersinus*) were not generally discriminated until about 1908." *E. chersinus* does not range north of New Jersey.

STROBILOPS LABYRINTHICA Say   *Labyrinth Snail*

SIZE: Minute, less than ⅛ inch high and about as wide.
DESCRIPTION: The shell is dome shaped, dull, chestnut brown, the entire surface minutely sculptured with oblique, regular, rather prominent ribs. The lip is reflected and there are 2 or 3 white ridges (or lamellae) issuing from the aperture. The umbilicus is small and deep.

This shell has much the same shape as *Euconulus* but is easily distinguished by its surface sculpture, the thick, dull texture, the reflected lip, and the internal folds at the base. It is not uncommon under the bark of dead trees on Long Island and near Peekskill.

At the Englewood Yacht Club near Palisades, New Jersey, we once took a large number of *Strobilops* which turned out to be the species STROBILOPS AENEA Pilsbry. This species is lower than *S. labyrinthica*, has a sharper angle at the base, is dark brown in color and has a wide umbilicus. A quick look through an ordinary hand lens is sufficient to differentiate between these two species.

In addition to these species of the genus, there is a third one reported from Staten Island, STROBILOPS AFFINIS Pilsbry. We have never collected this species personally, but according to its author, it can be recognized by being "somewhat larger than *labyrinthica*, thinner." The internal folds on the base of this

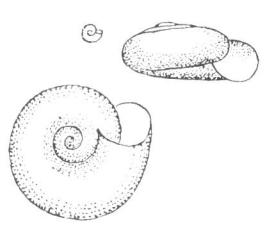

Hawaiia minuscula [×8]

Strobilops labyrinthica [×10]

Punctum minutissimum [×10]

species are almost equal in length, whereas in *S. labyrinthica* they are prominently unequal. The shell has to be broken open to observe this characteristic.

HAWAIIA MINUSCULA Binney *White Zonite Snail*

SIZE: Minute, less than ⅛ inch in diameter.
DESCRIPTION: The shell is depressed, with four narrow whorls. The color is white, hyaline, silky in appearance. The umbilicus is large, showing all the whorls to the apex.

The small, shiny, whitish shell with its large umbilicus is somewhat like *Vallonia* except for the lip, which is never reflected. For a time after we first found it, we took it to be a white bleached dead shell of some zonitid, until we realized that dead shells never have a gloss. The shell can be taken in leaf mold in Peekskill, Croton and in several places on Long Island.

The generic name for the species is due to an error. G. K. Gude, the author of the name, collected widely in Hawaii. When he found this shell, he did not realize that it was a mainland snail accidentally introduced into the Hawaiian Islands and gave it what he felt was a properly patriotic designation. When it was decided to remove these whitish snails from the genus *Vitrea*, which included many unallied forms, it was found that Gude's name was the earliest for the group and so the laws of priority prescribed that this misleading name be used.

PUNCTUM MINUTISSIMUM Lea *Tiny Mulch Snail*

SIZE: Very minute, less than 1/16 inch in diameter.
DESCRIPTION: The shell is light brown, somewhat shiny. The surface is finely striated with growth lines which are crossed by weak spiral cords. The umbilicus is wide and deep, the lip simple. There are about 4 whorls.

The tiny land mollusk, with its very appropriate name, can be found with other tiny shells in moist leaf mold in Peekskill, Long Island and elsewhere. In older local lists it appears under the name *P. pygmaeum*, an equally apt name. However, *P. pygmaeum* is a European shell.

Mollusks of Fresh Water

Large Pond Snail
LYMNAEA PALUSTRIS ELODES Say*

SIZE: ¾ to 1 inch high (frequently smaller), width less than one-half the height. DESCRIPTION: The shell is dull, dark brownish, tapering, elongated and quite thin and fragile. The last whorl is comparatively large, somewhat more than one-half the entire shell. The lip is thin, the columella white and twisted. The umbilicus is usually covered. Many specimens are malleated, that is they present the appearance of hammered copper.

One of the commonest members of the family Lymnaeidae, this species is found almost everywhere in ponds, lakes and slowly flowing brooks. Fine specimens can be taken at Alley Pond in Queens. The specimens from Annsville Creek and the Hudson River at Peekskill, which are subjected to the ebb and flow of tides, are considerably smaller.

LYMNAEA COLUMELLA Say   *Wide-mouth Pond Snail*

SIZE: Up to ⅞ inch high, ½ inch wide.
DESCRIPTION: The shell is light amber colored, fragile, moderately shiny. The aperture is about four times higher than the spire. The lip is thin, very fragile, the columella moderately twisted. The umbilicus is very tiny, usually covered by the columellar wall. The surface is sometimes malleated.

We have usually taken very large specimens of this shell on lily pads and submerged vegetation at Alley and Lilly Ponds in Queens in the early spring, when apparently they come to the shore to breed. The specimens taken in the summer and fall are generally much smaller.

This shell looks very much like the land snails of the genus *Succinea*. But it should be remembered that though both are air breathers, the habitat of *Lymnaea* is fresh water, whereas *Succinea* lives exclusively on land, although sometimes quite miry.

Ear-shaped Pond Snail
LYMNAEA AURICULARIA Linné

SIZE: Up to 1¼ inches high, nearly 1 inch wide.
DESCRIPTION: The shell is very fragile, very light amber or

---
* See List of Changes in Nomenclature, page iii.

dull green in color. The aperture is huge, flaring, ear-like in outline, the summit of the outer lip often reaching the same height as the very small, acute spire. The umbilicus is covered over; the columella very twisted above.

This shell, an accidental immigrant from Europe, is found in many parts of our country from California to New York and at least as late as 1907 was found still living in a lily pond in Prospect Park, Brooklyn. Though we have not succeeded in finding it in the New York area, we did collect it upstate at Cazenovia and chances are strong that this visitor will turn up again locally.

LYMNAEA HUMILIS Say *Humble Pond Snail*

SIZE: About ⅜ inch high, less than ¼ inch wide.
DESCRIPTION: The shell has a relatively high spire, the aperture being about equal to the spire in height. The color is brown to yellowish, dully shining. The umbilicus is a mere chink that is almost covered by the slightly twisted, pale columella. The outer lip is thin, unreflected.

The last of the lymnaeids is usually smaller in this area. It looks like a young *palustris,* from which it can be distinguished by the much narrower whorls. It is common in the ponds and lakes near Peekskill, where at times it congregates on the shores in tremendous numbers.

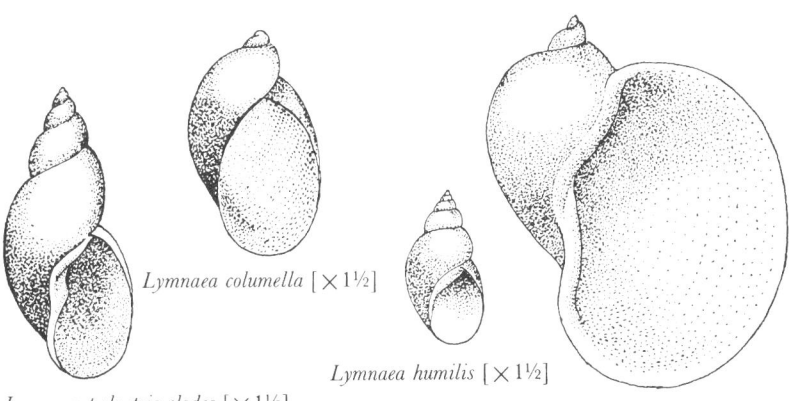

Lymnaea columella [× 1½]

Lymnaea humilis [× 1½]

Lymnaea palustris elodes [× 1½]

Lymnaea auricularia [× 1½]

This shell is frequently called *Fossaria humilis modicella* but in a recent work on the lymnaeids of the world, the author maintains that *modicella* is of doubtful value as a subspecific name, since it cannot be distinguished from true *humilis*. Furthermore, the characters upon which the generic name *Fossaria* was constructed are minor characters, unworthy of the rank of genus.

This region has two large species of Planorbidae or wheel or ram horn snails. Both belong to the genus *Helisoma*. They resemble each other a good deal, but can be distinguished without much trouble. Before we discuss these shells, it might be well to orient ourselves so that we can properly interpret the unusual form of these snails, which is, as the popular name implies, very much like a wheel. Holding the shell upright, the aperture above and facing away from us, we observe that the right side has scarcely three high, keeled whorls, and a deep umbilicus in the center; the left side has at least four very much flattened whorls, separated one from the other by a very lightly impressed suture. It takes but a little imagination to realize that actually the left side constitutes the spire, and the right side the base of a shell that has been simply telescoped together, like an old fashioned collapsible drinking cup, until all the whorls are in the same plane.

HELISOMA TRIVOLVIS Say   *Three-whorled Rams Horn*

SIZE: ¾ to 1 inch in diameter, about ½ or less in width.
DESCRIPTION: The shell is disklike, showing all the whorls above and below; the color is pale yellow, brownish or chestnut, the lip sometimes reddish and, in mature specimens, somewhat thickened within. In specimens from some brooks in Long Island near Valley Stream, the shell is not evenly round, but has several angled areas.

This species is found in shallow water clinging to submerged sticks, leaves, and other debris or is sometimes found walking on the muddy bottom. It particularly favors the underside of floating bits of wood, where it can frequently be collected in large numbers. It is found in practically every pond or slowly-moving stream in the New York area. The tiny stream flowing through the Brooklyn Botanic Garden has a large and prolific colony.*

---

* Another *Helisoma* species is described in the Supplement of Additional Species, page 137.

HELISOMA ANCEPS Menke *Keeled Rams Horn*

SIZE: Just under ½ inch in diameter, ¼ inch wide.
DESCRIPTION: The shell is disklike, the center deeply concave above and below. The whorls are strongly keeled on the left, the spire, side. Color yellowish to dark brown.

This *Helisoma* is easily separated from *trivolvis*, for on both sides, spire and base, it is considerably more excavated, the spire on the left side forming a deep pit, like an inverted cone, whereas in *trivolvis* the same area is quite shallow and flattened. In addition, *anceps* is smaller and the left side of the aperture is distorted by the end of a sharp keel which bounds the inverted spire. The habitat is the same as *trivolvis*, but it is unusual to find both species in the same body of water. We have taken *anceps* in Hempstead Park Lake and in many ponds on Long Island and around Peekskill.

Anceps has enjoyed two recent re-baptisms. In older books it is called *Planorbis bicarinatus* Say, but that was before the genus *Planorbis* was limited to very flat shells from Europe. Then it was found that Conrad had called the shell *P. antrosus* some years before Say used his name, and hence the Conrad name was in use for a time, until Dr. Pilsbry discovered the earliest available name to be that of the German conchologist Menke. Hence, through strict application of the laws of priority, the name is now, and will probably remain, as given above.

PLANORBULA JENKSII H. C. Carpenter *Jenks' Rams Horn*

SIZE: Little more than ¼ inch in diameter, frequently less; only ⅟₁₆ inch high. DESCRIPTION: The shell is flat, disk-

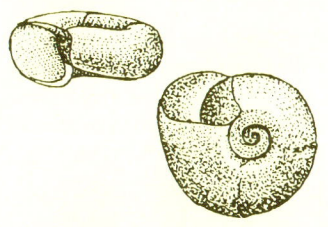

Helisoma trivolvis [× 1]

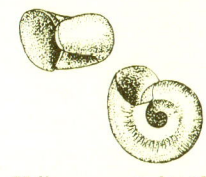

Helisoma anceps [× 1]

like, reddish to brown; about 4 whorls, the last one twice as wide as the others. The umbilicus is wide, showing all the whorls. The lip is thin, unreflected, but with a thickened ridge just within. About ⅜ of an inch inside the aperture there are 4 white teeth.

We have found this small wheel snail in many of the small ponds and lakes about Peekskill and on Long Island. Unlike the shells of the genus *Helisoma*, this shell has the umbilical region to the left side when the shell is held upright with the aperture away from the observer. The teeth within the aperture can be seen from the outside through the thin shell. It is a very pretty little snail that lives in forest pools containing masses of wilting vegetation. When these pools vanish in late summer, many *Planorbula* can be found in the dried beds.

GYRAULUS PARVUS Say *Tiny Rams Horn*

SIZE: Small, ⅛ to ¼ inch in diameter, ¹⁄₁₆ inch high.
DESCRIPTION: The shell is disklike, thin, reddish or grayish brown, black when alive. The 3 to 4 whorls are rounded at the periphery, the umbilicus is open.

This very small planorbid, with its nondescript coloration, is a common snail in every pond or slowly flowing brook, where it can easily be collected from the underside of bits of floating wood, lily pads, and rotting leaves. It can most easily be taken by a vigorous shaking of the fronds of aquatic plants. We have it from Lakes Beechwood and Mohegan near Peekskill, from Hempstead State Park, Kensington Pond, and Valley Stream on Long Island.*

PROMENETUS HUDSONICUS Pilsbry *Hudson Rams Horn*

SIZE: Very small, about ⅛ inch in diameter, very flat.
DESCRIPTION: The shell is shaped like a minute lens, very depressed, fragile. The periphery is sharply keeled, the umbilicus is open, the color grayish amber, but much darker when alive.

About the same size as *Gyraulus parvus*, this snail can readily be distinguished by its reddish brown color when taken alive,

---

* A related species, *Micromenetus dilatatus*, is described in the Supplement of Additional Species, page 137.

and, most prominently, by the distinct, rather sharp keel along the outer edge. It is also noticeably flatter. It is common in Wallace Pond near Camp Smith in Peekskill, as well as in Lake Mohegan and many other lakes and ponds in Westchester, Long Island and northern New Jersey. It too lives on submerged bits of timber and on the fronds of aquatic plants.

It was first recognized as a different species in 1934. The type was taken from a small woodland swamp near the Poughkeepsie Rural Cemetery. This snail was frequently confused with *Menetus exacuus,* and appears under this name on many older lists. However, the latter is a shell that is usually found further west.

PHYSA HETEROSTROPHA Say   *Tadpole Snail*

SIZE: ⅝ inch high, ⅜ inch wide, but usually much smaller.
DESCRIPTION: The shell is ovate (egg-shaped), the sinistral aperture more than one-half of the shell; smooth, shiny, transparent, pale yellow to brownish; the lip is smooth, the columella twisted. The body whorl is much larger than the spire.

The *Physa* are the common tadpole snails found in practically every body of fresh water, and are commonly seen in fish tanks. They are undoubtedly the most common fresh water snails in the New York City area. They can easily be recognized by the sinistral aperture, that is, the aperture is to the left of the observer when the shell is held upright. It is easy to recognize and to collect, but the weakness of the fragile lip calls for careful handling if perfect specimens are to be procured.

Although we have collected shells around Peekskill and on Long Island that have been assigned by several authorities to a

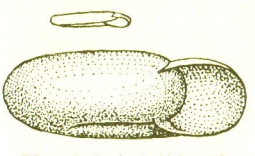
Planorbula jenksii [×5]

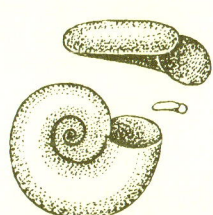
Gyraulus parvus [×6]

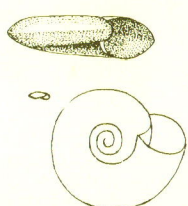
Promenetus hudsonicus [×6]

number of species, it seems that in the New York area there are actually only two species, *heterostropha* and PHYSA ANCILLARIA Say. The latter is distinguished from the former by the much shorter spire and more angular general outlines. According to Dr. William J. Clench of the Museum of Comparative Zoology of Harvard, *heterostropha* is the *Physa* of still bodies of water whereas *ancillaria* lives in streams and rivers.

Thriving *Physa* colonies live in the Brooklyn Botanic Garden, the ponds in Prospect Park, Bronx Park and Forest Park, in Alley Pond and Baisley Pond in Queens and in practically every body of fresh water that manages to persist through most of the summer drought.

APLEXA HYPNORUM Linné   *Polished Tadpole Snail*

SIZE: About ½ inch high, ⅜ inch wide, but usually smaller in this region. DESCRIPTION: The sinistral shell is brilliantly polished, smooth, thin, elongate, yellow, quite transparent. The body whorl is just about one-half of the entire shell. The lip is fragile, unreflected.

This close relative of *Physa* can be distinguished by its elevated spire and brilliant lustre. It is by no means as common as *Physa* and is best collected by straining the mud of small stagnant pools, especially in the spring and early summer. We took it from such a pool in the woods near Peekskill.

This is one of our handsomest fresh water snails because of its highly polished, transparent shell, graceful outline and bright yellow color. Dr. Gould, the great American conchologist, wrote of them as follows: "It is quite interesting to keep a number of them in a vessel of water and to observe their motions and habits. The manner in which they open their mouths and display the lingual organ, the manner in which they rise to the surface and open the air cavity, into which its structure permits no water to enter, and, above all, the beautiful and unaccountable manner in which it glides along, will never fail to excite astonishment."

FERRISSIA FUSCA C. B. Adams   *Fresh Water Limpet*

SIZE: Very small, ¼ inch long, ⅛ inch wide.
DESCRIPTION: The shell is cup-shaped like a small limpet.

It is very thin, pellucid, light brown at the edges, darker at the summit. The base is oval in outline, the apex slightly elevated and blunt.

Because of its diminutive size, this shell is easily overlooked. We found scores of specimens inside a dead *Anodonta* valve in Lake Mohegan near Peekskill, but it took close looking to detect them. It is also reported from a pond in Flushing, Queens, and near Huntington on Long Island.

FERRISSIA PARALLELA Haldeman is a similar shell. However, the base is not oval but has the two longer sides straight and more or less parallel. The color is light green, the apex higher and more pointed. We found this species on flat-bladed marsh grasses in a tiny woodland pond near Annsville, Westchester County, New York.

With *Ferrissia* we end our discussion of the air-breathing shells that have learned to live in fresh water. They can all be easily recognized by the fact that they do not possess an operculum, a door-like structure attached to the foot of the animal that serves to protect it when its soft parts are all withdrawn into the shell. The rest of the local fresh water snails possess this useful organ, usually made of a horn-like material. These snails breathe by means of gills and hence do not have to come periodically to the surface to replenish their air supply. They vary in size from the truly large *Viviparus malleatus* to the tiny *Amnicola limosa* and are found in most bodies of fresh water.

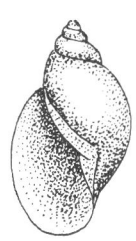

Physa heterostropha [×2]

Aplexa hypnorum [×2]

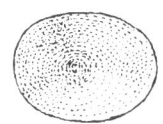

Ferrissia fusca [×3]

GONIOBASIS VIRGINICA Gmelin *Virginian Horn Shell*

SIZE: Little more than 1 inch high, ⅜ to ½ inch wide.
DESCRIPTION: The shell is turreted (very elongate), rather thin, smooth but not glossy. It is brown or olive in color; sometimes it has two revolving, reddish bands. The aperture is rather narrow, bluish within, somewhat drawn out at the base. The lip is thin, somewhat wavy at the margin. There is a dull, horny operculum.

The pleurocerids, of which this shell is a member, are very greatly developed in the rivers of Georgia, Tennessee, and Kentucky. This is the only representative found in our area. It can be collected in New Jersey in the North Branch of the Raritan River.

VIVIPARUS MALLEATUS Reeve *Chinese Mystery Snail*

SIZE: 2½ inches high, 1½ inch wide.
DESCRIPTION: The shell is roundly ovate, olive green, the spire brownish. The aperture is bluish, the lip raised in a dark, narrow, reflected band. The earlier stages of this lip appear as thin, irregularly placed, longitudinal ribs. Juvenile shells have a row of hair-like processes at the periphery; mature shells show three faint, spiral lines of weakly impressed dots where these hairs were lost. The umbilicus is narrow; the operculum transparent, reddish amber, horny, ear-shaped.

Viviparus, as its name indicates, does not lay eggs, but gives birth to live young. Frequently on cleaning this snail, we have discovered as many as sixty small embryos provided with a tiny operculum and ready to sally forth into their watery world. When less mature, these embryos consist of small square globs of albumen about 3/16 inch in size. A gravid snail transported to a suitable aquarium will readily give birth. The male is similar in general appearance to the female, but is smaller.

This *Viviparus* is not native to our country. It was probably accidentally introduced with exotic water plants and has since flourished in suitable lakes and ponds in many places on the northeastern coast. It has become a favorite aquarium snail, and at times whole barrels are offered for sale in Chinatown, where no doubt it becomes the ingredient of some delectable oriental dish. We have found magnificent specimens in Lilly Pond in Hollis, in Prospect Park Lake and the pond near the Japanese pavillion in Brooklyn, in Forest Park in Queens, in Clove Lakes on Staten Island and in the Saw Mill River opposite Nepera Park in Westchester County.

Common Mystery Snail
VIVIPARUS CONTECTOIDES Binney*

SIZE: 1½ inches high, about 1 inch wide.

DESCRIPTION: The shell is higher than wide, moderately shiny when not covered with a deposit, light olive-green, with 3 or 4 indistinct brown bands on the body whorl. The lip is thin, unreflected; a thin black line runs along the very edge. The aperture is less than one-half the entire shell; the operculum as in *V. malleatus*.

This snail looks superficially like *V. malleatus,* but it can readily be distinguished by its smaller size and the presence of the brownish bands. Hairs are not present.

It is prevalent in our country to the Mississippi River and is common upstate around Warrensburg and elsewhere, but we have never found it near Peekskill and it is not listed from Long Island. Recently, however, it was discovered in the lakes of Central Park. The presence of a colony here may be due to the activi-

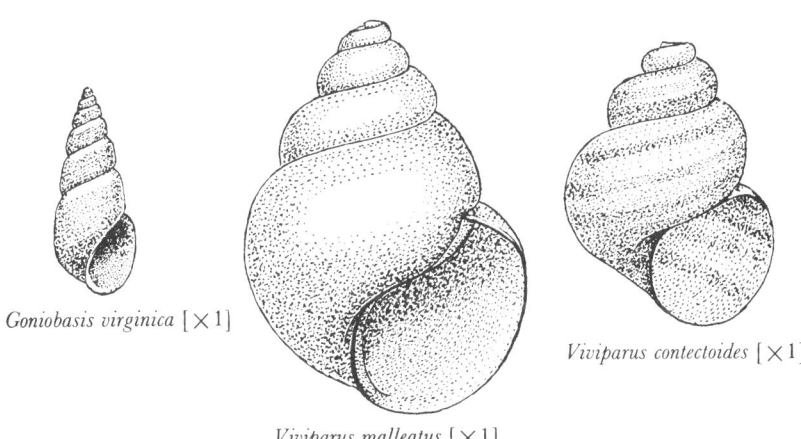

Goniobasis virginica [×1]

Viviparus contectoides [×1]

Viviparus malleatus [×1]

* See List of Changes in Nomenclature, page iii.

ties of an unknown snail collector, who was trying to establish this *Viviparus* in our city. His success is evidenced by the fact that huge numbers are flourishing near the boat house of the 72nd Street pond. However the water is apparently not the purest, since most of the older specimens are extremely eroded.

CAMPELOMA DECISUM Say   *Small Mystery Snail*

SIZE: About 1 inch high, ½ inch wide.
DESCRIPTION: The shell is thick, elevated, olive to yellowish-green. The sutures are shallow; the lip thin, not reflected. There is no umbilicus. The operculum is horny, honey colored.

This snail usually has 5 whorls, but the tip of the spire is invariably eroded, sometimes so badly that about half the shell is gone and the poor animal has to crowd into its greatly reduced domicile. This species is not uncommon locally. It is found in ponds and muddy streams, usually concealed under shelving banks or imbedded in an inch or two of loose mud. We found it on the banks of the Hudson at Croton Point State Park and elsewhere near Peekskill. There is a record of it from Riverhead, Long Island, and it can probably be found in many other bodies of water on Long Island.

Common Faucet Snail
BULIMUS TENTACULATUS Linné

SIZE: ½ inch high, ¼ inch wide.
DESCRIPTION: The shell is conic, rather thick, yellow and shiny when the dark outer layer is removed. The sutures are shallow, the lip is unreflected, the umbilicus barely perceptible. The operculum is corneous, same color as the shell.

In 1879 this immigrant from northern Europe was first detected in the Great Lakes. Since then it has spread rapidly, and in 1898 appeared in the Chicago water works in numbers large enough to clog the screens and issue forth at the faucets. According to an article in "Natural History" for February 1950, it is now common over much of the northeastern section of our country. We found it in large numbers in Canopus Creek at

Annsville, near Peekskill, and in many other streams and ponds in the same area.

AMNICOLA LIMOSA Say *Tiny River Snail*

SIZE: 3/16 inch high, 1/8 inch wide.
DESCRIPTION: The shell is bronze green or brown (yellowish or dirty white when dead). The 4 whorls are set off by deep sutures. The lip is unreflected, the umbilicus a mere chink.

This snail is one of the commonest operculates in our area and is taken in every suitable body of fresh or even brackish water. It is found on submerged plants and under submerged stones and sticks. It can be scooped in large numbers from sand or mud. We have taken this shell in Canopus and Annsville Creek near Peekskill and there are records from many bodies of water in Westchester and on Long Island.

VALVATA TRICARINATA Say *Three-keeled Valve Snail*

SIZE: About 3/16 inch in diameter, less than 1/8 inch high.
DESCRIPTION: The depressed shell is light tan or faintly greenish when alive. The 3 or 4 whorls are flattened, with three prominent keels, one each at the summit, the base and the center of the body whorl. The sutures are channeled, the aperture round, but somewhat modified by the terminations of the three keels. The umbilicus is deep, funnel-like.

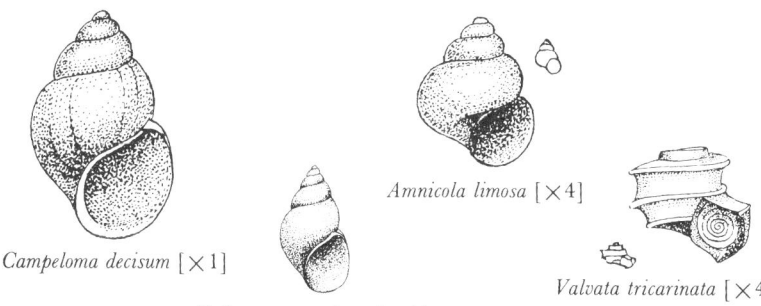

Campeloma decisum [×1]

Amnicola limosa [×4]

Bulimus tentaculatus [×1]

Valvata tricarinata [×4]

One of our most curious fresh water shells is this flattened species with its neatly sculptured, sharp and prominently raised triple keel, which makes it look like a tiny turban shell of tropic seas. It lives under stones and in the discarded shells of fresh water mussels, and can sometimes be dredged in mud.

We have found this shell only in Canopus and Peekskill Creeks at Annsville near Peekskill, and since none of the older lists mention it at all, we must assume that that is its nearest approach to the City. There are no Long Island records.*

Fragile Fresh-water Mussel
ANODONTA CATARACTA Say

SIZE: 4½ to 5 inches long, about 2½ inches high.
DESCRIPTION: The valves are long, elliptical, thin, shiny, inflated, and marked by uneven lines of growth. The color is greenish yellow or olive, usually with indistinct bands of brown. The interior is bluish white, somewhat shiny. There are no interlocking teeth at the beaks.

The river mussels that form so prominent a part of the molluscan fauna of the lime-rich rivers of the middle west of our country are here represented by three rather plain species. The first is the paper-thin, greenish, inflated shell that is found in many ponds and streams on Long Island. It is readily separated from its only relative, whose description follows, by the almost complete absence of teeth at the hinge, by its inflated shape, and by its extreme thinness. Older records report it from Baisley Pond and Kissena Park Lake in Queens and Prospect Park Lake in Brooklyn; we have found fragments in Alley Pond, and single valves in Central Park Lake at 72nd Street. It is common in Lake Ronkonkoma, and in Lake Mohegan near Peekskill.

Like all fresh-water mussels, *cataracta* lives half sunk in the mud or sand of the lake bottom, its lower portion well submerged, a small portion of the upper part extending beyond the mud and containing the open siphons which take in fresh water and food particles, and give off wastes. It can move, but only very slowly, by the use of its powerful, tongue-like foot.

As late as 1907 there exist reports of the presence of a closely related mussel in Prospect Park. This is ANODONTA IMPLICATA Say which differs from *cataracta* chiefly in its thicker shell and more yellowish color.

* Another *Valvata* species is described in the Supplement of Additional Species, page 137.

ELLIPTIO COMPLANATUS Dillwyn* *Filter Tank Clam*

SIZE: 3 to 3½ inches long, 1½ to 2 inches high. Larger elsewhere. DESCRIPTION: The shell is smaller, heavier, less inflated and much thicker than *Anodonta*. The outline is roughly quadrate, and the periostracum rough, brown and lusterless. The hinge of the left valve has a long, thin lamella and a rough pyramidal tooth directly behind. Both these structures fit into corresponding grooves on the right valve. The internal color varies from bluish-tinged white through warm peach and light salmon to light violet.

This is the commonest fresh-water mussel in our area, being present in fair numbers in most larger bodies of fresh water. Like

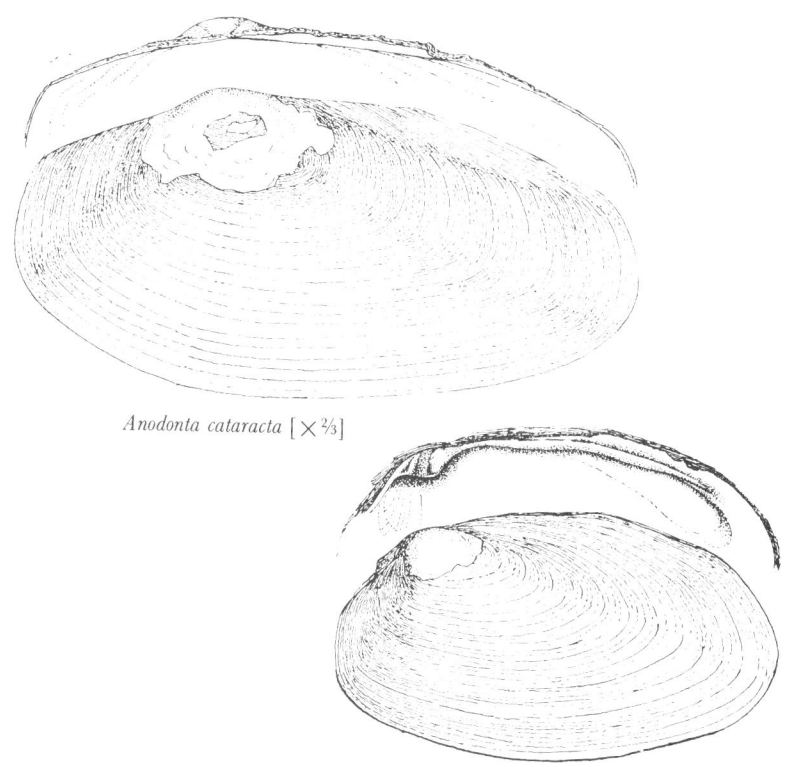

Anodonta cataracta [× ⅔]

Elliptio complanatus [× ⅔]

* See List of Changes in Nomenclature, page iii.

Anodonta it cannot reproduce successfully unless fish are present, since in the larval stages the clam must encyst itself upon such a host, where it leads a semi-parasitic existence for a short time. In a fishless stream this life cycle is broken and the clam disappears.

Elliptio is highly variable in shape and in older shell books enjoyed no less than seventy-four different and distinct names. Even a late work (1927) allots it eleven different sub-species. However, identification should not be difficult if due respect is paid to the few distinctive characteristics described above. We have found it near Peekskill, on Long Island and in the Saw Mill River near Yonkers.

This clam is used by fish fanciers as a natural filter to keep the water in home aquaria clean. Microscopic organic particles in the water are used as food by the clam, hence its popular name.

Pearl-bearing Mussel
MARGARITANA MARGARITIFERA Linné*

SIZE: About 4 inches long, 2 inches high.
DESCRIPTION: The shell is elongated, more pointed behind, rounded in front. The ventral margin is strongly arched in the center in mature shells. The periostracum is tarry black and quite smooth. The interior is bluish white, sometimes tinted flesh color in the center. The left valve has two stumpy teeth under the umbo, the right valve one that fits

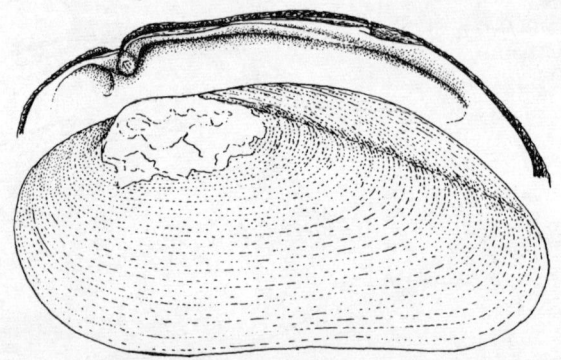

Margaritana margaritifera [× ⅔]

* See List of Changes in Nomenclature, page iii.

into them. There are no long ridge-like teeth on either valve as in *Elliptio*.

This mussel is the source of handsome, at times valuable, fresh water pearls (hence its name) and is harvested in several European countries for this purpose. We found it in the headwaters of the Hackensack River near Haverstraw in Rockland County. It is larger, somewhat more inflated than *Elliptio,* with which it lives together in this locality. In addition it has a darker and much smoother periostracum. These characteristics, as well as the arched ventral margin and absence of the long tooth-like ridges in the interior of the valves, set it off immediately.

This shell has a fantastically wide range, from Western Europe clear across the two continents to the Ussuri River in Siberia and in the islands of Japan. In our continent it appears in New England and Canada and the northeastern states, and again in the mountain rivers of California and the far west. It was difficult for our early malacologists to accept so vast a range for a species which is confined to rivers, and the shells in our country received various names. Nevertheless they can not be separated from the European and Asiatic specimens. The large range can probably be explained by the travels of the host fish which the young of mussels inhabit for a time.

SPHAERIIDAE *Fingernail Clams*

This is the family of fresh-water clams that includes the usually delicate, fragile shells, appropriately called "fingernail clams." They are frequently held to be the young of the true river mussel, which of course they are not. In color they vary from very pale buff through light snuff color to dark brown. The family contains three genera which are found in our area.

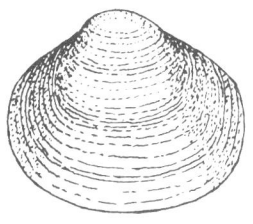

Sphaerium [×3]

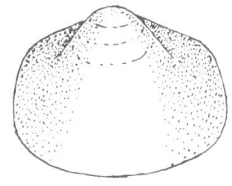

Musculium [×6]

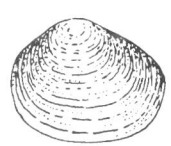

Pisidium [×6]

In the pond above the dam in Piermont, New York, and in Clove Lakes on Staten Island we found shells of the only true *Sphaerium* species in this area. These belong to the species SPHAERIUM RHOMBOIDEUM Say and are the largest members of the family Sphaeriidae found hereabouts. The adult shell is swollen, the ends truncated or cut off so that the shell appears somewhat four-sided or rhomboidal in outline. The surface is shiny, delicately marked with fine, rounded concentric wrinkles, the color on the outside being a rather dark brown but lighter and more reddish in the beak area. The interior is light bluish. Specimens are nearly ½ inch long, and about ⅜ inch high. The young of this species are much less inflated, thinner, and considerably lighter in color; however, they do have the characteristic rhomboidal shape.

The other species of fingernail clams, which belong to the genera MUSCULIUM* and PISIDIUM, are very difficult to identify, the experts themselves being far from any agreement. Hence we shall limit ourselves to a discussion of the generic characteristics and a listing of the species reported from the New York area.

Members of the genus MUSCULIUM* can be distinguished by their comparatively large size (but usually smaller than *Sphaerium* proper), much thinner texture, lighter color, and chiefly by the beaks, which are quite prominent and rise into a cap-like cover. The usually delicate shell is up to ⅜ of an inch in diameter, but most of our species are no more than 3/16 of an inch. The following species have been reported from the New York area, but few scholars will be bold enough to name any of them definitely: *M. jayanum* Prime, *M. partumeium* Say, *M. securis* Prime, *M. truncatum* Linsley. Almost any body of fresh water will support fair numbers of this genus.

The genus PISIDIUM includes shells that are much smaller, sturdier and relatively more inflated than the *Musculium*. They differ most noticeably from *Sphaerium* and *Musculium* in the position of the beaks, which are always to one edge, thus giving a lop-sided, triangular outline to the shell. They attain ⅛ inch in size. The species of this genus are even harder, if possible, to name than *Musculium*. The following are our local species: *P. abditum* Haldeman, *P. variabile* Prime, *P. noveboracense* Prime. These occur in the same habitats as the other two genera.

---

* Read *"Sphaerium"* instead of *"Musculium."* The distinction between the two names no longer holds.

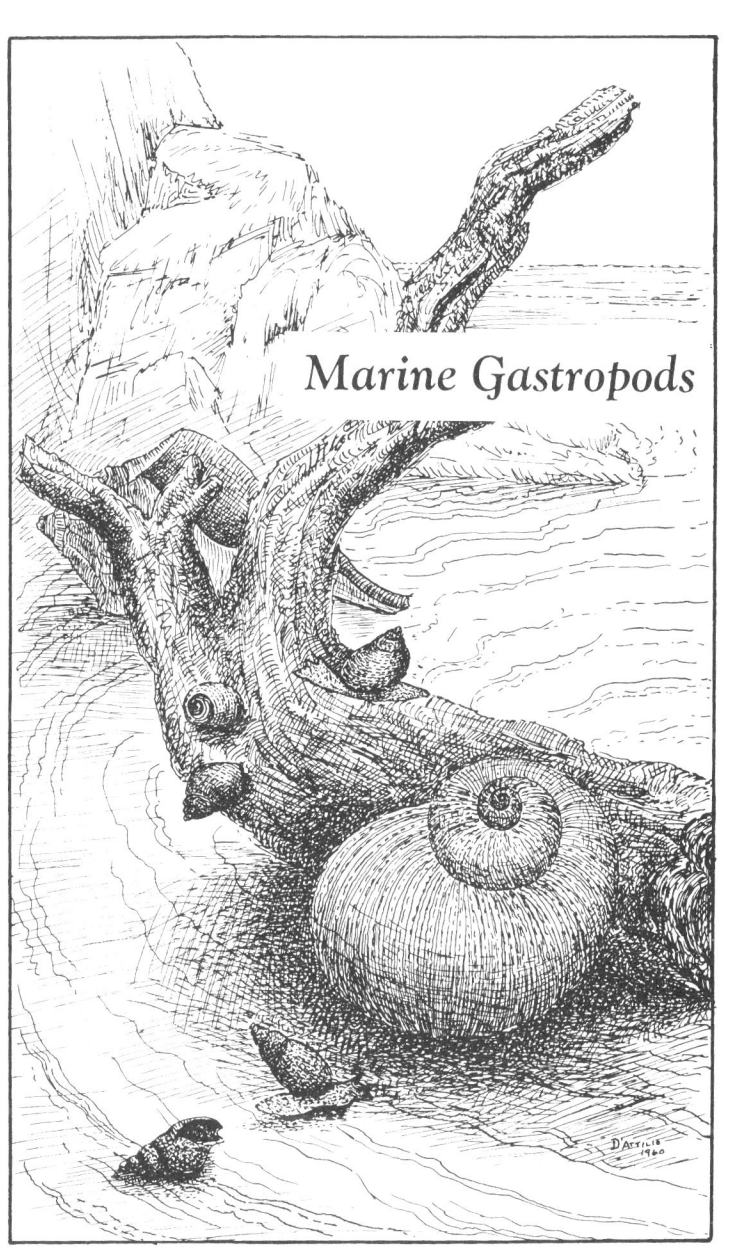

Marine Gastropods

ACMAEA TESTUDINALIS Müller   *Chinese Hat Shell*

SIZE: ⅝ inch long, little more than ½ inch wide, ⅜ inch high. Larger in New England. DESCRIPTION: Shell shaped like an inverted cup, oval in outline, rather thin. The exterior is greenish white with dark brown, broken lines radiating from the summit. The interior is shiny, the central portion frequently brownish; the inside margin bordered with alternating small brown and white areas.

The description of our marine snails appropriately begins with the rock limpet, for of all snails, it possesses the simplest shape, merely a flattened, shallow cone. This shape is thought to be the archetype of the primitive univalves from which all properly spiral snails have descended. *A. testudinalis* is the only true limpet of our shores and it can be found clinging tightly to rocks just at the low tide mark. We have taken it at Bayville on the North Shore and it can probably be found elsewhere where the coast is rocky. Live shells can be collected by inserting a knife between the foot of the animal and the rock, but one must move fast, otherwise the snail, alarmed by the presence of an enemy, clamps down hard and can not be removed without damage to the shell.

Actually the acmaeas found in the New York area belong to the subspecies *A. testudinalis fergusoni* Wheat. This subspecies, first described from the North Shore of Long Island, is very close to the typical *testudinalis*. The chief difference is its smaller size. In

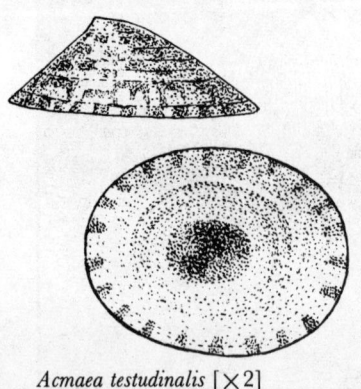

Acmaea testudinalis [×2]

Epitonium rupicola [×2]

addition it is more convex than the typical, more rounded and less elongate, and when alive, is apparently a more active animal than its sluggish northern relative.

Sometimes this snail makes its home on the broad blades of eel grass. Specimens from such a habitat have nearly parallel margins, are much smaller and more fragile and have a roughly checkered pattern. There is some doubt as to the true status of this form, some malacologists holding it to be a distinct and valid species called *Acmaea alveus* Conrad, others making it a subspecies and still another group calling it a mere varietal form. It may be found living in the vicinity of typical *testudinalis*.

EPITONIUM RUPICOLA Kurtz *Purple Wentletrap*

SIZE: 1 inch high, just less than ½ inch wide at the widest portion. DESCRIPTION: The shell is high, elongate, cone shaped. The surface has numerous narrow, blade-like axial ribs, which are worn off in many specimens. The color is brown or banded brown and light brown, with large white areas. The aperture is nearly circular and surrounded by a thickened, white lip.

This beautiful wentletrap snail, the commonest in our area, can be found readily at Ward Point near Tottenville, Staten Island. It has also been found at Bayville and many other localities on Long Island. It is rather fragile for its genus, and dead specimens frequently suffer from large breaks. The animal gives off a violet dye when it is irritated. It was until recently known as *E. lineatum* Say.

Epitonium multistriatum [×2]

Epitonium angulatum [×2]

Epitonium humphreysii [×2]

EPITONIUM species *Wentletraps*

The other wentletraps are small, white shells that are occasionally found dead on our beaches. Their outstanding feature is a number of white, gleaming axial ridges or costae reaching from the spire to the base. In addition to *rupicola,* the following three species occur in our range, not necessarily in New York City: E. MULTISTRIATUM Say, which has many low, narrow, closely-spaced ribs; E. ANGULATUM Say, which has fewer ribs, characterized by sharp angles or shoulders as they cross the sutures, the whorls being slightly separated at this point; and E. HUMPHREYSII Kiener, which is difficult to separate from *angulatum,* but is proportionately narrower, has rounded rather than blade-like ribs and a much less developed angle at the whorl shoulder. We have found isolated specimens of each of these species at Rockaway Beach and dredged some in Great South Bay in about 8 feet of water.

POLINICES HEROS Say *Hero Moon Snail*

SIZE: Usually 2½ inches long, 2 inches wide. Fully mature specimens grow as large as 4½ or 5 inches. DESCRIPTION: The shell is inflated, globular, the body whorl huge; the aperture is large, egg shaped. The color is ashen brownish, sometimes with large bluish areas, covered by a thin yellowish periostracum. The columella is covered by a thin enamel layer, the interior dark brown, brilliantly polished in live animals. The operculum is horny, amber color, roughly semi-circular.

The Natica or moon snail (to which group *Polinices* belongs) has such a huge foot that one wonders how it can ever fully retract into the shell. This, however, the creature accomplishes by ejecting a large quantity of water, which is apparently the source of its vast inflation. The animal is predaceous, living upon clams and snails, including its own brother moon snails. It gains entrance by drilling a beautifully countersunk hole with its file-like radula and a queer acid-secreting gland. Many shells bearing such a hole can be found; in others the hole is only half broken through, the victim somehow having made an eleventh hour escape. In many shells of the fossil Calvert formation near Washington, D. C., which is about 50 million years old, this same type of hole appears, evidence of primordial tragedies that long pre-dated the advent of shell collectors. The Staten Island Museum has in its collection a Venus clam shell that bears two such holes. It is difficult to guess how this came about, since *Polinices* will not attempt to drill dead shells.

P. heros can be dug up alive at low tide by opening the small hillocks it throws up while plowing its way through the wet sand. The egg case of this snail is a jar-like structure of grains of sand and mucus, in the walls of which the tiny eggs are imbedded. These "sand collars" are commonly found in the spring on most beaches.

This species used to be called *Natica heros,* but the true Naticas are now considered to be those that have a stony or calcareous operculum, whereas *Polinices* have a horny one. These opercula, amber colored and translucent, are frequently found free in large numbers. Recent shell books place *heros* in the genus called *Luna-*

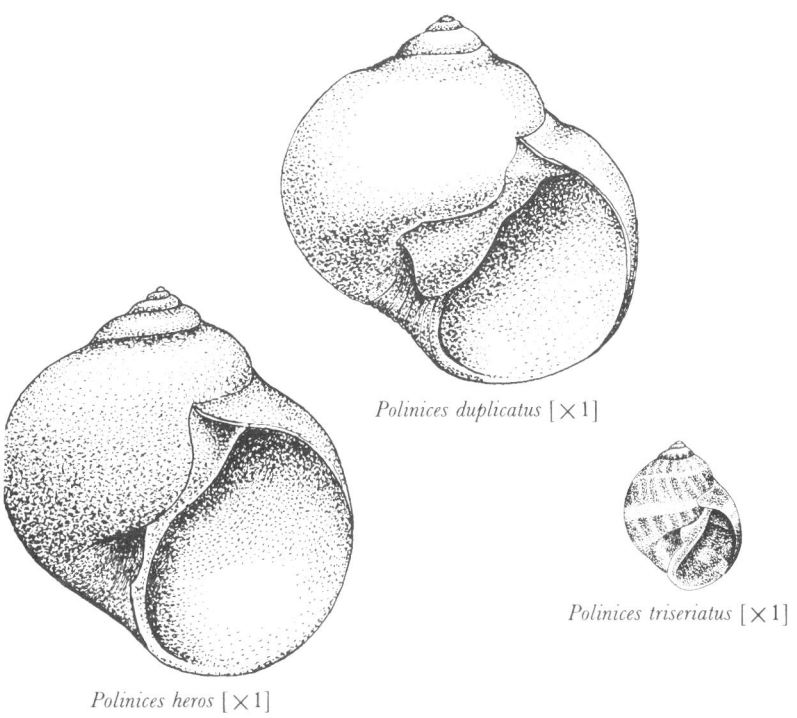

Polinices duplicatus [×1]

Polinices triseriatus [×1]

Polinices heros [×1]

tia, but there do not seem to be sufficient grounds for giving this subgenus full generic rank.

POLINICES DUPLICATUS Say *Callused Moon Snail*

SIZE: 2 inches high, 2 inches wide.
DESCRIPTION: The shell is solid, depressed globular, heavy. Many specimens have the spire produced in a solid, widely pyramidal peak. The lip is thin and sharp. The umbilicus is almost, but not quite, covered by a heavy brown or purple callus. Color and operculum are the same as in *P. heros*

This moon snail, with much the same habits and habitat as *heros,* is smaller, heavier and can easily be distinguished by the heavy brown or purple callus. It is not so rounded, being widest near the base. It is less common than *heros* on our beaches, but completely replaces the latter as one proceeds south below Cape Hatteras.

POLINICES TRISERIATUS Say *Three-lined Moon Snail*

SIZE: ¾ to ⅞ inch wide, ¾ inch high.
DESCRIPTION: The shell is small, solid, dull, yellowish white, sometimes with three series of brown elongate spots arranged in horizontal lines. The periostracum is thin, brown. There is a shiny white callus near the small, deep umbilicus. The operculum is thin, horny.

One June day several years ago, the New York Shell Club went to Cold Spring Harbor on the north shore of Long Island, known since 1870 as a famous collecting area for marine mollusks. Here almost immediately we found a sand collar, the egg case of a moon snail, which did not look quite like the one we were used to finding in Rockaway. It was much smaller and had the "neck of the jar" less elevated, so that it looked somewhat like a lip-less tea pot. A little later we found the artisan of this interesting object, a small, dully shining snail with a yellowish shell. The typical shell of this species has the three bands that give it its specific name. Such specimens are frequently found at Rockaway, but only as dead shells generally covered with hard,

[stony *Hydractinia* concretions, evidence that the shell has served as the home of a hermit crab.]

On a later occasion, this time in May, on a visit to Bayville, not far from Cold Spring Harbor, we collected large numbers of this snail, as they plowed diligently through the sand in shallow water, leaving a distinct trail behind. We followed the twisting trails to the end and there dug up the snails. Some of these shells were very faintly banded, enough to establish their undoubted relationship to the distinctly decorated shells from Rockaway. The animal is of a very pretty lemon or bright orange color.*

CREPIDULA FORNICATA Linné   *Slipper Shell*

SIZE: Up to 1½ inches, height very variable.

DESCRIPTION: The slipper shell is elongate cup-shaped and has a dull white base color, variously and handsomely decorated with interrupted wavy longitudinal markings of a light chestnut color. The apex is turned sharply to one side and affixed to the shell wall. The interior is shiny light brown with dark spots and blotches. It has a large white septum or shelf, the free end of which is widely sinuous.

The height of the shell varies, depending upon its habitat: some, taken from the carapace of a horse-shoe crab, are extremely flat; others, growing on rounded pebbles and stones, are very arched. On the muddy, stone-free beaches of Nantucket, these snails, hard put to it for a solid place to rest, attach themselves to one another till an eight-inch "stick" of Crepidulas is formed, each shell clinging tightly to the one underneath. Because of this habit of clinging to objects in large numbers, *Crepidula* has been regarded as a menace to young oysters, which are literally suffocated under the weight of bulky slipper shells. It has been accidentally introduced into England where it is causing considerable uneasiness among oystermen. We may have learned to deplore the presence in our country of the blustering English sparrow, but it is well to bear in mind that the mother country did not escape unscathed in this unwitting faunal exchange.

This snail is an inhabitant primarily of quieter bays and inland waters where it appears in astounding numbers. Huge

---

* Two related species, *P. immaculatus* and *Natica pusilla*, are described in the Supplement of Additional Species, page 137.

windrows of these shells are found near Huntington, Long Island, and along many points on the shore of Long Island Sound. It is far more common in Coney Island than in Rockaway, thus showing that it does not prefer the outer beaches.

Its startling Latin name is related to the noun *fornix* which means an arch, a rather apt name for this arched slipper shell.

CREPIDULA PLANA Say *Lady's Slipper Shell*

SIZE: 1¼ inch long, ¾ inch wide.
DESCRIPTION: The shell is white, flat, the outer surface unevenly wrinkled, the inner surface with a thin, usually raised, septum.

This shell, the lady slipper, is related to the previous one, as can be seen by the white diaphragm, but is so distinct as to cause little confusion. It is white, wrinkled and always extremely flat. It lives inside the aperture of dead Naticas, Busycons, and other large snails. The shape varies greatly, depending upon its immediate habitat. It is found quite commonly on the outer beaches, somewhat less frequently in protected bays.

CREPIDULA CONVEXA Say *Convex Slipper Shell*

SIZE: ½ inch long, ¼ inch wide.
DESCRIPTION: The shell is cup-shaped, thick, very concave. The outer surface is faintly wrinkled, ashen-brown in color with stripes or dots of reddish brown. The septum is deeply set, with a cavity leading deep into the apex. The free edge of the septum is concave. The apex is directed straight back.

This, the last of the *Crepidula* tribe found in the New York area, is somewhat less common than the other two. It can be readily distinguished from small specimens of *Crepidula fornicata* by the character of the apex, which is acute and centered, whereas that of *C. fornicata* is strongly turned to one side and usually united with the margin of the aperture. In addition, the diaphragm is more deeply situated, and its free edge, instead of being wavy, has a simple curve. *C. convexa* is more likely to be taken alive in sheltered bays, where it occupies much the same habitats as *C. fornicata*. In Jamaica Bay, near Beach 116th Street in Rockaway, we once collected 25 fine specimens on the outside of a discarded beer bottle.

CRUCIBULUM STRIATUM Say *Cup and Saucer Limpet*

SIZE: 1⅛ inches in diameter, ½ to ¾ inch high.

DESCRIPTION: The shell is cup-shaped, conical, the apex twisted and with numerous radiating lines. The septum is white with a brown spot at the base, about one-third of it attached to the shell wall; the rest is free and sharply curving. The color is wax yellow, dull; the interior shiny light brown, becoming darker toward the apex.

This species is like a *Crepidula* with the septum modified to an almost triangular shape, attached by one of its edges to the side of the shell. It has a deep groove that ends in an acute tip. Its common name, cup and saucer limpet, is descriptive, the larger shell being the saucer, the deep septum serving as the cup. It is quite rare on outer beaches. When it is found it is usually dead and much the worse for wear. The only living specimen we ever collected was attached to one valve of the ocean scallop, *Pecten magellanicus*.

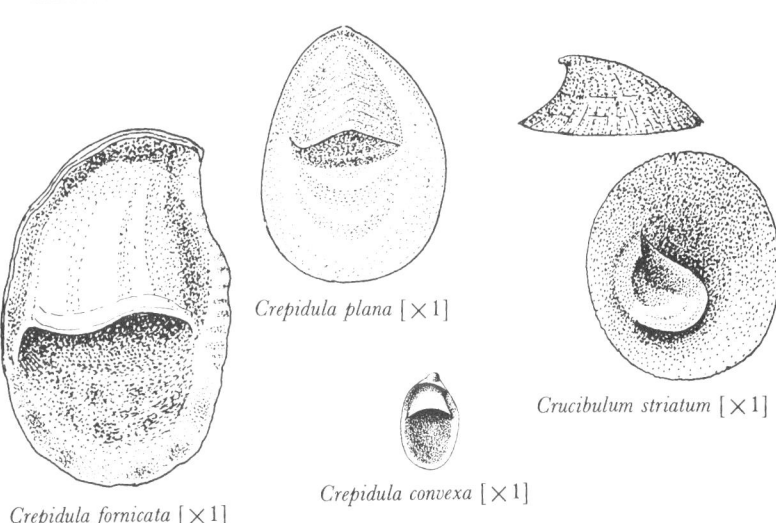

Crepidula plana [×1]

Crucibulum striatum [×1]

Crepidula convexa [×1]

Crepidula fornicata [×1]

LITTORINA LITTOREA Linné   *Common Periwinkle*

SIZE: ¾ to 1 inch high, about as wide. In favorable areas specimens may reach 1½ inches in height. DESCRIPTION: The shell is heavy, solid, chunky, with very shallow sutures. The color is dull brownish or ashy, occasionally decorated with very narrow rings of red, black or smoky brown. The aperture is smooth, chocolate brown, the columellar lip white. The outer lip is sharply beveled. There is a thin, dark, horny operculum. No umbilicus. As the shell ages, it becomes more elongated and a depression appears under the suture of the body whorl.

This European snail first appeared in Nova Scotia about 100 years ago and advanced steadily southward to southern New Jersey. It quickly became the commonest species of rocky shores, where it is found in bewildering numbers. At Hunters Island, Pelham, the specimens are smaller than in Jamaica Bay, but just as common. Lazy collectors can buy pounds of fine specimens in many fish markets in Italian neighborhoods where, as in England, it finds a ready sale as a special delicacy. It is simply cooked in salt water and the contents extracted with a pin. The young of this species are covered with strong revolving ribs which soon disappear as the snail matures.

LITTORINA IRRORATA Say   *Marsh Periwinkle*

SIZE: ¾ to 1 inch in height, ½ to ¾ inch wide.
DESCRIPTION: The shell is thick, elongate conical, body whorl of adults with regularly spaced, shallow longitudinal grooves. The color is grayish white with a bluish tinge in younger specimens, with blotches or short streaks of dark purple or reddish brown on the spiral ridges. Aperture pear-shaped, a brownish orange area on the columella. No umbilicus; operculum thin, brownish.

Sometimes on the beaches in New York one can find a dead, broken shell that differs from *L. littorea* by being higher, more pointed and with faint traces of blotches or short streaks on the last whorl. These pitiful specimens are apparently the only signs still remaining of this snail, once abundant on these shores. Some believe that it has vanished because of the increasing coldness of our off-shore waters. It is still abundant in Maryland and further south, but in our area there is no definite historical record.*

LITTORINA SAXATILIS Olivi *Rough Periwinkle*

SIZE: ¼ to ½ inch high, ¼ inch wide. At Montauk Point they grow to ⅝ inch. DESCRIPTION: The shell is comparatively thin, yellowish, ash, greenish or orange in color. The surface is marked with shallow revolving lines. The sutures are deep, the spire smaller than the body whorl. Umbilicus wanting, lip thin, operculum horny.

This shell is easily separated from the previous one by its smaller size, yellowish or brownish gray color, and rounded whorls with deep sutures. At Pelham Bay we once collected several which were beautifully marked with one or more spiral white bands, and others of an orange gray color with dark purple bands. This variation of color gave rise to a number of named varieties. The male shell is smaller than the female and has a somewhat longer spire in relation to the body whorl. It is found clinging to rocks, piling, and jetties. It can be collected easily in Jamaica Bay, Cold Spring Harbor and elsewhere at low tide.

LITTORINA OBTUSATA Linné *Round Periwinkle*

SIZE: ¼ to ½ inch high, about as wide.
DESCRIPTION: The shell is thick, dull, rather smooth, round. The spire is very small, the body whorl about ¾ of the entire shell. The color is variable: yellowish, brownish, some-

Littorina littorea [×1]

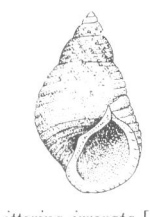

Littorina irrorata [×1]

Littorina saxatilis [×1]

Littorina obtusata [×1]

* Recently a living specimen was reportedly found in Jamaica, near Rockaway Beach.

times greenish or black. Some specimens have broad spiral bands. Umbilicus wanting, operculum bright yellow or orange.

L. obtusata can easily be identified by its flattened spire which gives the shell a globular appearance. This small shell is usually found under *Fucus* seaweed on rocks. It is common near the Riis Park entrance of the Marine Parkway Bridge in Rockaway and on other rocky situations. The male shell is smaller than the female and has a somewhat more elevated spire.

LACUNA VINCTA Montagu *Lesser Periwinkle*

SIZE: ¼ inch high, ⅛ inch wide.
DESCRIPTION: The shell is small, thin, high with a pointed spire. It is dingy white or more often purplish brown; there are four darker chestnut-colored spiral bands. The outer lip is sharp, thin, the columella white, flattened, with a smooth crescent-shaped groove ending in an umbilicus.

The adult sometimes grows to about one-half inch in height but we have always found the tiny young to be much more common. We have taken large numbers as they clung to red seaweed that was left on the beach at Rockaway by the receding tide. This species is most easily identified, even in bleached, dead shells, by the flattened columella with the excavated crescent groove.

UROSALPINX CINEREA Say *Oyster Drill*

SIZE: 1¼ inches high, ⅝ inch wide.
DESCRIPTION: The shell is fusiform (shaped like an old spinning spindle), with 9 to 12 rounded axial ribs which are more widely spaced on the body whorl. These ribs are crossed by shallow, narrow spiral ridges. Aperture with a short, backward-bending canal at the base. Color usually bluish or purplish brown or ashen; aperture sometimes pinkish brown. Operculum horny, amber colored.

The menacing oyster drill turns out to be a harmless looking, rather handsome shell. This bane of the oystermen does its work like the Naticas by clamping itself firmly to its victim with its orange-cream colored foot and then bringing its radula into play. It bores a tiny hole into the shell from which it drains the soft body of the clam. It can be found at low tide on pilings or under pebbles and rocks, especially at the bay or Sound beaches.

In some colonies the color of the specimens is variable. At Ward Point on Staten Island, we have found large numbers of orange and white specimens, some nicely banded. The pure white shells, extremely handsome objects, have been named *Urosalpinx cinerea atkinae* by Silas Wheat, the Brooklyn conchologist. However, these shells intergrade so completely with typical specimens, that the sub-specific name is probably useless.

EUPLEURA CAUDATA Say   *Thick-lipped Oyster Drill*

SIZE: ½ to 1 inch long, ½ inch wide.
DESCRIPTION: The shell has a narrow, almost enclosed basal canal and strongly angled, crown-like whorls. There are about eleven transverse ribs, two of which are enlarged at the periphery of the shell into wing-like varices, giving a peculiar flattish appearance to the entire structure. The ribs are crossed by equidistant lines. The aperture has a thick outer lip, provided with small whitish teeth within. The color is dark mahogany, covered by a grayish blue layer. The aperture is bluish white or brown.

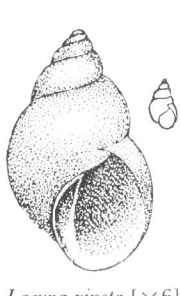

Lacuna vincta [×6]

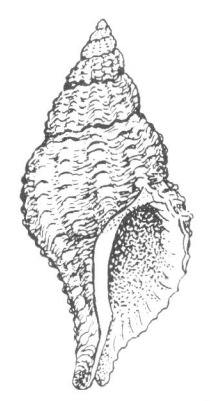

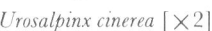

Urosalpinx cinerea [×2]

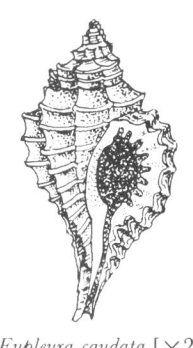

Eupleura caudata [×2]

This species, like *Urosalpinx,* belongs to the family Muricidae or rock shells, and it is not an unworthy representative of such a highly ornamental group. Also known as an oyster drill, it occupies much the same habitats as *Urosalpinx,* but it is a good deal less common.

THAIS LAPILLUS Linné*   *Common Purple*

SIZE: 1½ inches high, about ⅝ inch wide.
DESCRIPTION: The shell is pointed at both ends (fusiform), thick, solid, not shiny. In color it varies from white through yellow to chocolate, often prettily banded. It is sculptured with coarse ridges, which are sometimes smooth and sometimes armed with raised, concave scales that make the whole shell rough to the touch. The oval aperture is usually brown or yellow and has a sharp but thickened outer lip which is provided with a row of blunt teeth at a little distance within. It has a horny, elliptical operculum.

This shell, so common everywhere on rocky shores in New England, comes no nearer New York than the eastern tip of Long Island, as far as Hither Hills State Park. It is carnivorous and lives on the barnacles that cover so many tidal rocks.
T. lapillus is a relative of the famous snail of the Mediterranean Sea, from whose crushed bodies the Phoenicians extracted an indelible purple stain, the royal Tyrian purple.

ANACHIS AVARA Say   *Narrow Dove Shell*

SIZE: ⅜ to ½ inch high, about ¼ inch wide.
DESCRIPTION: The shell has an elevated, pointed spire and a surface covered with minute revolving lines that are interrupted by a series of ten to fifteen upright, smooth waves or rounded ribs. The suture is well impressed and somewhat scalloped by these folds; the lip is thickened on the outside and bears a series of very fine elongated teeth just within the aperture. When fresh, the shell is light-straw colored, blotched with various shades of reddish brown.

This small, graceful relative of the lovely Columbellidae or dove shells, is easily found on the north shore of Long Island and

---

* See List of Changes in Nomenclature, page iii.

especially in the bays that open into the Sound. We have taken it in large quantities on Shelter Island and in Shinnecock Bay and it is reported to be abundant at Cold Spring Harbor. Older records say it abounded as well on Staten Island and elsewhere within the city limits, but whether through a lack of diligence in searching, or because of a change in our local fauna, we must confess that years of hunting have yielded us only two imperfect specimens within the city limits, one from Rockaway and one from Pelham Bay.

There is a smaller variety of this shell with finer and more numerous longitudinal ribs that bears the name *Anachis avara similis* Ravenel. The two specimens that we found within the city limits seem to belong to this form.

Its stout structure and comfortable shape make this shell a favorite of the younger generations of hermit crabs, and in areas north of us it is a common sight at low tide to see this shell scooting busily over the wet sand, propelled by the active limbs of these small crustaceans that have it in temporary possession.

MITRELLA LUNATA Say — *Lunar Dove Snail*

SIZE: Small, 3/16 to 1/4 inch high, 1/8 inch wide.

DESCRIPTION: The surface is smooth, somewhat shiny, reddish brown or faun colored with two or three series of crescent-shaped whitish spots. The color varies; frequently brown unspotted shells or almost completely white ones

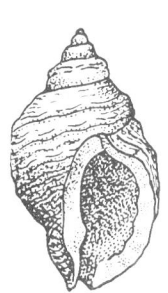

Thais lapillus [×1]

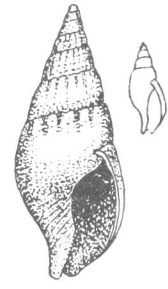

Anachis avara [×3]

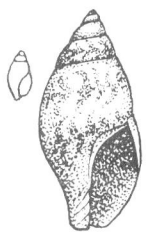

Mitrella lunata [×6]

are found. The outer lip is slightly thickened, dark brown in color and toothed along the inner margin.

This small, shiny, prettily colored shell is the only other member of the Columbellidae or dove-shell family in our area. It is quite common at Rockaway and elsewhere, where the best specimens can be collected as they cling to bits of seaweed that have been thrown on the shore. It is easily dredged in shallow water in Great South Bay and elsewhere. The shape is similar to *Anachis* but is dumpier and not so graceful. It is considerably smaller.

Recently at Fire Island a colony of *Mitrella* was found which are much darker in color and completely unmarked by white crescent marks. We are strongly tempted to give this a special varietal name, but so far have bravely resisted the urge to add another Latin name to this much overnamed family.

ILYANASSA
~~NASSARIUS~~ OBSOLETUS Say*   *Mud Basket-snail*

SIZE: ¾ to 1 inch high, ¼ to ⅜ inch wide.

DESCRIPTION: The dark reddish brown shell is ovular in shape, with a moderately high spire, blunt apex (usually eroded in mature specimens) with a distinct suture; the shell is covered with a granular sculpture, formed by numerous revolving lines broken by axial growth lines and numerous oblique folds. The columella is deeply arched and overspread with a shining dark violet enamel that reaches far into the aperture. A bluish white band is frequently present, but is most often visible only within the aperture. Both color and sculpture of the shell cannot be seen until the layer of black mud and a greenish mold-like plant growth is removed. There is a horny operculum with smooth edges.

This is one of the most commonly found snails in mud-bottomed bays. As the tide goes out, the mud flats are literally covered with countless specimens, usually much eroded and quite unprepossessing, that move about restlessly leaving a well-defined trail behind. As the tide stays out and the sun begins to beat down more strongly, the snails try to bury themselves in the soft muddy sand. Any dead flesh is quickly buried under feasting

---

* See List of Changes in Nomenclature, page iii.

Nassas. The animal is grayish and quite active, twisting swiftly in its shell and showing a most un-snail-like agility.

A colony living on the rocks at the Riis Park end of the Marine Parkway Bridge has the shell sculpture so well preserved that the specimens look almost like a different species. The shells living at Canarsie are unusually large but very eroded.

Three-lined Basket-snail
NASSARIUS TRIVITTATUS Say

SIZE: ⅝ to ¾ inch high, ¼ to ⅜ inch wide.
DESCRIPTION: The shell is greenish or yellowish white, and is covered with a network of raised lines which give it a cross-barred effect. The whorls are flattened above with a conspicuously shouldered appearance at the suture. The three lines for which this snail is named are usually lacking in the specimens found in our area. The operculum is corneous and has one half of its edge shallowly serrated.

This dweller in clean ocean sand is rarely taken alive in the Rockaways, but dead shells, all neatly bored on the last whorl, are quite common. They occur in particularly large numbers at Far Rockaway. Live specimens are found at Bayville, Cold Spring Harbor and other localities on the North Shore of Long Island where the beaches face the open Sound. Here the snails live shallowly buried in the sand in about 2 feet of water.

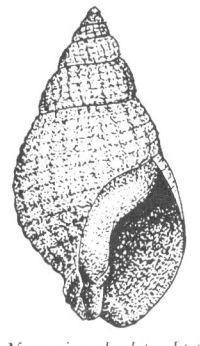

Nassarius obsoletus [×2]

Nassarius trivittatus [×2]

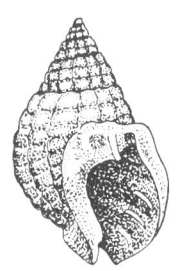

Nassarius vibex [×2]

NASSARIUS VIBEX Say *Southern Basket-snail*

SIZE: ⅝ inch high, ⅜ inch wide.
DESCRIPTION: The shell is short and chunky, the surface roughened by waving folds and revolving lines. There is a heavy splash of gleaming enamel reaching from the aperture to cover most of the adjacent part of the body whorl. The outer lip is thickened and has several shallow, toothlike protuberances within. The color is bluish-brown or ashen, with a few horizontal brown lines at the aperture. The columella is bluish white, the aperture brown.

These basket-shells, which are so common further south, are quite rare in our area. Dead shells are occasionally found in Rockaway and on the North Shore of Long Island. There are reports of live specimens taken at Huntington in 1870 and at Lloyd Neck in 1899. We recently found a live colony at the eastern end of Shinnecock Bay, living in a very restricted area in about four feet of water, as well as a rather extensive colony off The Sunken Forest of Fire Island in Great South Bay.

This shell was once named *N. fretensis* by Perkins, and some malacologists insist that our northern form is sufficiently different from the southern shells to be labeled *Nassarius vibex fretensis*. But that is a matter of opinion.

BUCCINUM UNDATUM Linné *English Whelk*

SIZE: Up to 4 inches high, 1½ inches wide.
DESCRIPTION: The shell has six convex whorls, decorated with about 12 obliquely undulating ribs covered by fine spiral ridges or, in some specimens, by prominent raised lines. When taken alive, the animal has a velvety periostracum, a pleasing tone of yellow in the aperture and an operculum that does not completely close the aperture.

At Rockaway, Long Beach and Jones Beach, we have occasionally found a grayish mass about the size of a large fist, composed of small, transparent, tough membranous sacs, each of which contained a number of small unborn shells. This is the egg case of the British whelk, badly worn mature specimens of which are much less frequently collected. The only live specimens of this snail we ever saw were "collected" in a Sheepshead Bay fish store. It is a common shell of Massachusetts, but we have found it to be quite rare in the New York area.

In England it is a very popular article of diet, but it is not usually so used in the United States. The two Sheepshead Bay specimens referred to above were found in a basket of *Busycon* whelks.

COLUS ISLANDICUS Gmelin*   *Iceland Whelk*

SIZE: Up to 2¾ inches long, ¾ inch wide.

DESCRIPTION: The shell is fusiform, with a graceful, rising spire and a body whorl that is more than half the length. Its entire surface is covered with regularly spaced, moderately deep revolving lines which, at their terminations, give the thin-edged lip a finely crenulated appearance. There is an open, extended columellar canal which is bent gently backwards. In life the shell is covered with a horn-colored, velvety periostracum. The aperture is polished, porcelain white.

The finding of a shell as rare as this one usually constitutes a red letter day for the patient beachcomber. In all we have found six more or less badly battered specimens, all thickly encrusted by the coral-like animal remains of *Hydractinia*, an animal that

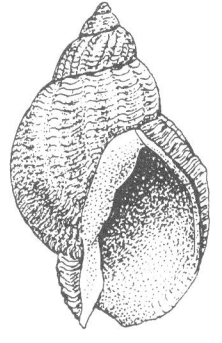

Buccinum undatum [×½]

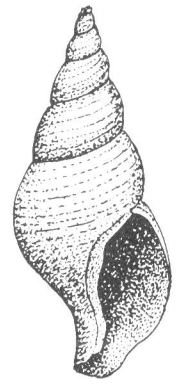

Colus islandicus [×1]

* See List of Changes in Nomenclature, page iii.

lives commensally with a hermit crab. It is a deep-water shell which within its range, north of Cape Cod, is best collected "ex pisce"—from fish stomachs. The specimens we find here are strays, brought down by currents and widely ranging hermit crabs. Hence its rarity should not surprise us.

COLUS STIMPSONI Mörch is just as rare. It is a larger shell (5 inches by 1½) and thinner, but has the same outline as *C. islandicus*. We found one specimen of this shell on Fire Island. It is much more common in New England.

COLUS STONEI Pilsbry is an extinct species, being found only in the post-Pliocene deposits on the coast of New York and New Jersey known as the Jacobs Sands.* The shell differs from the other of its genus in having a dumpier shape, deep sutures and prominent spiral lines. However, usually only fragments of these fossils are found washed onto the beaches. It is not uncommon on the New Jersey shore and our find of one specimen at Rockaway constitutes the only New York record.

Channeled Whelk
BUSYCON CANALICULATUM Linné

SIZE: Very large, 8 inches long, 5 inches wide.
DESCRIPTION: The shell is thin, yellow or faun colored, covered in life with a hairy periostracum. The outer lip is thin, the body whorl strongly shouldered, the outer edge of these shoulders prettily beaded in young specimens, the beads flattening out as the animal matures. There is a deep channel at the suture. The operculum is large, rough, grayish brown.

This and the following species are the largest marine snails in our area. They live in comparatively shallow water in Long Island Sound and in the bays of the south shore of Long Island. Dead shells are not uncommon at Rockaway and Coney Island.

The egg case of *Busycon*, a common object at certain seasons on our beaches, consists of a series of flattened membranous capsules, each about the size of a half-dollar, attached at one end to a tough cord-like structure. Each capsule is filled with numerous unborn shells. The capsules of the knobbed whelk have an upright edge as in a coin; those of the channeled whelk have an edge that comes to a single keel.

---

* *Colus stonei* has also been found as a fossil on Gardiners Island and in a well at Westhampton Beach, Long Island.

BUSYCON CARICA Gmelin *Knobbed Whelk*

SIZE: Very large, 5 to 9 inches high, 5 inches wide.
DESCRIPTION: The shell is pear-shaped, heavy, ashen color but with axial streaks of brownish purple when young. There is no periostracum. The operculum is like that of the preceding species. The sutures are shallow and the outer edge of the body whorl bears a row of low knobs. The color of the aperture is bright brick red or gleaming yellow.

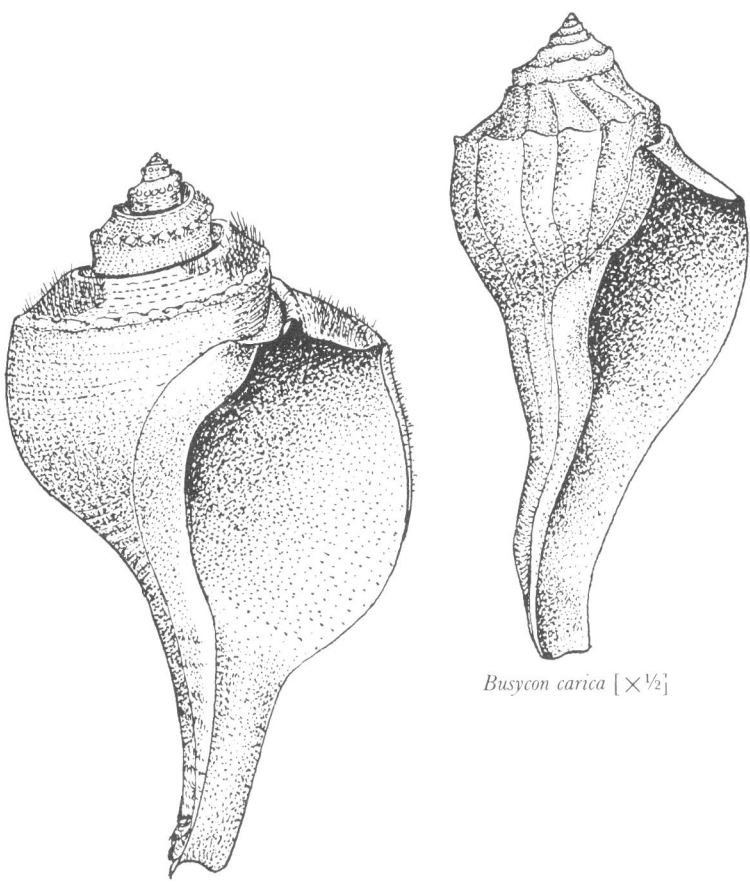

Busycon carica [× ½]

Busycon canaliculatum [× ½]

This species is common in shallow water and specimens are frequently thrown upon the beaches by storm waves. Both species are commonly sold in Italian fish markets, for the feet of these snails form the base of a delicious dish called *scungili*. These animals are carnivorous.

HAMINOEA SOLITARIA Say — *Fragile Bubble-shell*

SIZE: About ⅜ inch high, less than ¼ inch wide.
DESCRIPTION: The shell is barreled, the last whorl completely enveloping all the others, and the spire resting in a deep depression. The shell is thin, bluish white, very fragile, with a glassy texture.

This snail lives on the mud in very shallow water. We have it from Centerport, Cold Spring Harbor, Oyster Bay and Glen Cove. When the scoop net brings up one of these animals, it looks like a small blob of grayish flesh. In a moment, however, the mantle is withdrawn and the little bubble shell is revealed. It looks a good deal like a miniature *Bulla* of tropic seas, but is much thinner and more transparent.

MELAMPUS BIDENTATUS Say — *Common Marsh Snail*

SIZE: ⅜ inch high, 3/16 inch wide.
DESCRIPTION: The shell is brownish, smooth and shiny and occasionally, especially in young specimens, provided with rather wide revolving bands; the body whorl is very large, comprising about five-sixths of the entire shell. The other whorls are flattened and form a blunt spire. The aperture is long and narrow, the inner lip covered with enamel and provided with two tooth-like folds. The outer lip is thin and has one to four revolving ridges within.

This small air-breathing snail lives in huge colonies on marshy land that is covered and uncovered by the tide, on sedges and in wet places under boards and washed-up debris. We have found it along the marshy shores of Jamaica Bay and on Staten Island and in every similar situation in our area.

OVATELLA MYOSOTIS Draparnaud
Oval Marsh Snail

SIZE: Just over ¼ inch high, ⅛ inch wide.

DESCRIPTION: The shell somewhat resembles *Melampus* but it is smaller, more slender, and has a higher spire; the body whorl is proportionately smaller. The aperture is ovate and has three white folds or teeth on the inner lip. The outer lip is thin and without teeth. It is usually shiny and brown, but violet, bluish and white specimens are not rare.

This close relative of *Melampus*, until recently called *Phytia myosotis*, is also an air breather and can be found in the crevices of wooden piles and jetties near the high water mark. At Centerport we found some between the rocks that form the breakwater. It has also been taken in Jamaica Bay, Cold Spring Harbor and other points in Long Island.

The minute shells of the sea shore are even more numerous in specimens and species than their similar-sized relatives of the land and fresh water. Unless dredging operations are undertaken, they are not always easy to find. Our experience tells us that they can best be collected, at least at Rockaway, in the early morning with a retreating tide and very gentle waves. At such a time a thin line of fine shell debris is gently deposited at certain areas along the beaches, and this debris, collected in a bag, dried and examined at leisure under a weak magnifying glass yields these tiny shells in good numbers. We also collected large num-

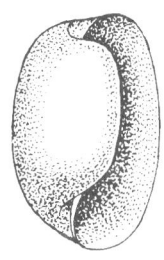
Haminoea solitaria [×3]

Melampus bidentatus [×3]

Ovatella myosotis [×3]

bers by dredging in 10 feet of water in Cold Spring Harbor and Great South Bay.

These shells actually constitute the bulk of local marine species and are most interestingly designed and shaped. However, we will have to limit ourselves to only a few that we have found most common and whose identification is easiest. Since even specialists in the field are not always agreed as to the determination and nomenclature of these Lilliputians, we frequently have had to be satisfied when we traced them down to their proper genus. Most of these shells are under 5 mm. in length.

BITTIUM ALTERNATUM Say   *Alternate Bittium*

This is probably the commonest of these tiny shells. It is quite narrow, high-spired and covered with a dense granular network of fine lines and grooves. When alive, the color is gray or slate and the young are usually very dark. We find numerous specimens of a dull brick color in beach drift. In Long Island Sound and in Great South Bay it lives in vast numbers on eel grass in shallow water.

CERITHIOPSIS GREENII C. B. Adams   *Green's Snail*

Smaller and narrower than *Bittium,* this shell can be most easily distinguished by the fact that the sutures are less apt to cause a constriction in the shell outline, whereas in *Bittium* each whorl is convex and sharply separated from its neighbor by the deep-seated suture. In addition there is a short basal canal, which is lacking in *Bittium.* It lives in sand and mud, under stones and shells, but seems to be far less common than *Bittium.**

SEILA ADAMSI H. C. Lea   *Adams' Snail*

This small shell, about ½ inch in length, is most easily recognized by the fact that the entire surface is covered by sharp, evenly spaced spiral cords like the threads of a tiny wood screw. In life it is chocolate brown or light yellowish, but dead specimens are whitish in color. We have found many specimens in the drift at Rockaway and in dredgings from Great South Bay.

---

* Another *Cerithiopsis* species is described in the Supplement of Additional Species, page 137.

TRIPHORA PERVERSA NIGROCINCTA C. B. Adams
Reverse Horn Shell

This is most easily distinguished from the other minute shells by the fact that the aperture opens to the left instead of to the right. It is from this feature that it derives its very fitting "middle name." Specimens attain the size of *Cerithiopsis* and are found in the same habitats.

ODOSTOMIA species
Rice Snail

These small shells are mostly tiny white specimens with flattened whorls and scarcely impressed sutures. They possess a tiny fold or tooth at the columella and the surface is frequently incised with shallow revolving lines. We have found them at low tide clinging to dead shells, stones and seaweed. Identification of species is not easy—even experts disagree at times—but the genus is readily recognized by the presence of the columellar tooth. The following species are reported from our area: *O. trifida* Totten, *O. impressa* Say, *O. seminuda* C. B. Adams, and *O. bisuturalis* Say. The first of these is common at Howard Beach in Jamaica Bay.

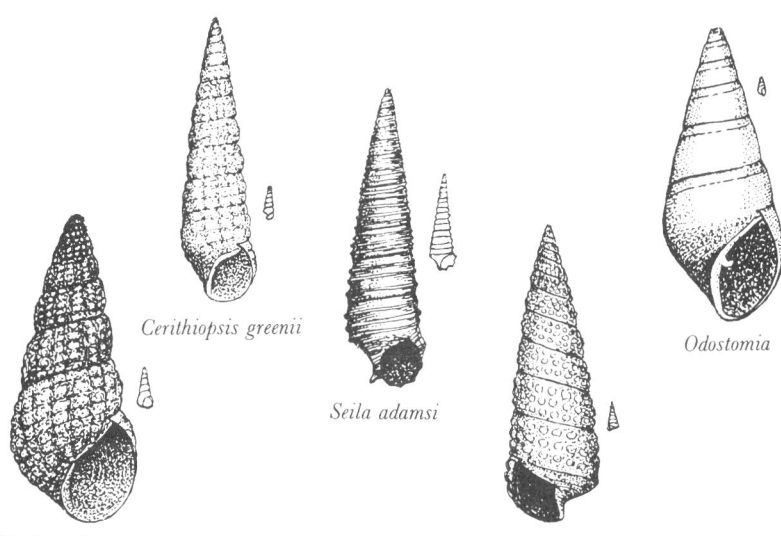

Cerithiopsis greenii

Odostomia

Seila adamsi

Bittium alternatum

Triphora perversa nigrocincta

TURBONILLA species — *Elongated Rice Snail*

Closely related to *Odostomia* are these slender, elongated many-whorled shells of which a few species occur in the New York area. On the beaches we have usually found them dead and with several whorls missing. We have dredged them in Great South Bay near Fire Island. They have more whorls than *Odostomia* but lack the columellar tooth. The most peculiar feature is that the earliest whorls wind to the left, while the rest turn to the right. The following species are said to occur here: *T. interrupta* Totten, *T. winkleyi* Bartsch, and *T. areolata* Verrill. These shells are sometimes seven mm. high.

HYDROBIA MINUTA Totten — *Minute Seaweed Snail*

When the broad-leafed seaweed *Ulva*, collected in the shallow portions of the quiet salt water bays, is placed in a pail of fresh water, this little shell, if present, detaches itself and sinks to the bottom. It is less than four millimeters high. It is thin, smooth and has a blunt apex and well-impressed sutures. When fresh, the shell is yellow-brown and finely translucent. It may be quite common in salt marsh pools. Dr. Gould had his troubles with this species and says of it: "This shell is so plain as to present no striking mark of distinction, and it is consequently not easy to describe it." It differs from *Odostomia* by its deeper sutures and the absence of the columellar tooth.

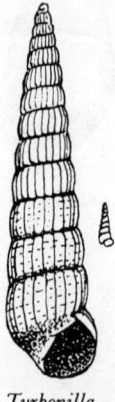

Turbonilla

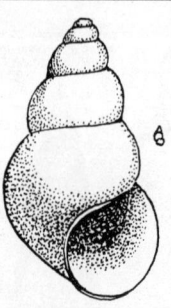

Hydrobia minuta

Mangelia cerina

Mangelia plicosa

MANGELIA species *Notch-lip Snail*

These small fusiform shells are white and heavily ribbed. There is a small slit, or sinus, at the top of the body whorl. They belong to the family Turridae, which is one of the largest groups of marine snails and correspondingly difficult to understand. They rarely grow larger than ½ inch. The commonest in our area is MANGELIA CERINA Kurtz & Stimpson, found on the beaches of Shelter Island and elsewhere on Long Island Sound. It has about 10 heavy, well-shouldered ribs. MANGELIA PLICOSA C. B. Adams has about 12 such ribs, rounded at the summit rather than shouldered. We dredged a few specimens in Great South Bay in about 8 feet of water. This species frequently has a gray or reddish brown color.

RETUSA CANALICULATA Say *Channeled Bubble Shell*

This shell looks like a tiny *Bulla* or bubble shell. It is cylindrical in shape, white and shining when fresh, with a very slightly elevated spire. The whorls are finely grooved or channeled at the top, the last whorl covering practically the entire shell. We have found dead specimens of this shell to be quite common in beach drift at Rockaway and have dredged live specimens from shallow water at Cold Spring Harbor and in Great South Bay. It reaches about four mm. in length. Until recent years this species was included in the genus *Acteocina*.

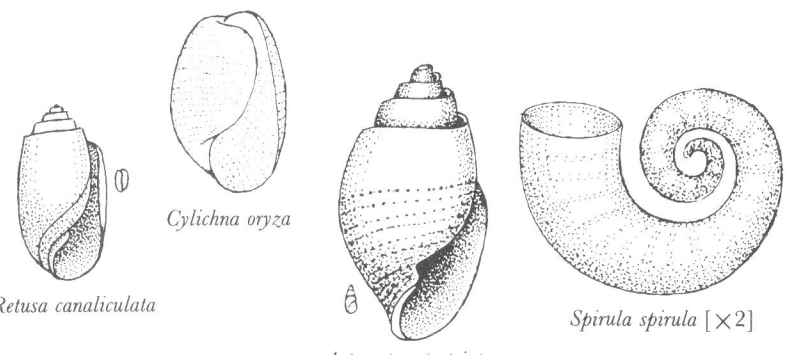

Retusa canaliculata

Cylichna oryza

Acteon punctostriatus

Spirula spirula [×2]

CYLICHNA ORYZA Totten *Minute Bubble Shell*

This tiny shell is the size and color of *Retusa*, but can immediately be distinguished by the position of the spire which, instead of being elevated above the large body whorl, is depressed and located in a shallow pit on the summit. This characteristic and its more rounded outline give it even more the appearance of a tiny bubble shell. Found with *Retusa*.

ACTEON PUNCTOSTRIATUS C. B. Adams *Punctured-line Bubble Shell*

This species is readily distinguished from the two preceding species by the fact that the body whorl covers a much smaller portion of the shell, rarely exceeding three-quarters of its entire length. The spire, though low, is more prominent; the whorls have a fine channel at the suture. Most distinctive though, are the ten to fifteen deeply incised revolving, punctured lines on the lower portion of the body whorl. This feature, easily seen under a hand lens, immediately serves to distinguish the species. Found with *Retusa* and *Cylichna*, which it resembles in size and color.

SPIRULA SPIRULA Linné *Sea Rams Horn*

SIZE: About 1 inch in diameter.

DESCRIPTION: The shell is pearly white, wound into two or three loose coils that are not in contact. The coils are sculptured with a series of constrictions, each one of which corresponds to the presence of an internal partition so that the entire shell is divided into chambers whose only contact is by means of a small tube or siphon placed to one side.

We once collected three specimens of this very beautiful little spiral shell on the beach at Hither Hills State Park, where it had probably been deposited by an outrunner of the Gulf Stream. This is the internal shell of a small cephalopod (like a squid). These shells are commonly washed ashore on tropical and subtropical beaches of the world. Only one species is recognized.

The species *Assiminea modesta* and *Caecum pulchellum* are described in the Supplement of Additional Species, page 137.

Marine Pelecypods

Clams and mussels usually make poor beach specimens, chiefly because it is difficult to collect specimens of many species with both valves in place. However, we are not ashamed to admit single valves to our collection if we have no doubles from the same locality. Professional malacologists frequently attach much value to such specimens because they serve, if for nothing else, to indicate the presence of the species in an area.

In the course of our discussions we have sometimes had to refer to the anterior or front portion and posterior or hind portion of the valves. Since to the unpracticed eye it is difficult to distinguish the front from the back end of a clam, we give the following hints as guides. (See also the diagram on page 123.)

The mollusk should be held so that the umbos or beaks point upward and the edges of the valves face the person holding the shell. If now the shell is turned so that the ligament located near the beaks faces the observer, then the part toward the observer is the posterior, the part away from the observer the anterior end. From this position it is an easy matter to determine the right and left valves. When the valves are opened and separated, a faintly impressed line will be seen in the interior running more or less concentrically with the basal margin and some distance from it. This line is called the pallial line and represents the point at which the mantle of the living animal is attached to the shell. At the posterior portion of the shell, as determined by the position of the ligament, this line is bent into an angle of varying depth. This angle is known as the pallial sinus, and since it is always located in the posterior portion of the shell, it enables one to orient the shell properly when the ligament is hard to find or even completely absent. To sum up: when the shell is held beak upward, the ligament and the internal pallial sinus are toward the posterior portion of the shell, the right valve is to the right, the left valve to the left of the observer.

The ligament is a brown, tough cartilaginous structure of varying size—sometimes very large as in *Venus,* sometimes very small as in *Tellina*—whose function is to act as a spring and keep the valves apart. The force opposing this tendency is exerted by the one or two adductor muscles, the points where these are attached being apparent as scarred areas of varying size and shape in the interior of the valves at the terminations of the pallial line. It should be borne in mind that when the clam is alive and closed, it is under muscular tension. Dead shells spring apart.

MARINE PELECYPODS · 73

SOLEMYA VELUM Say *Awning Clam*

SIZE: Up to one inch long, ½ inch wide, frequently smaller.
DESCRIPTION: The valves are very thin, fragile, elongate. The periostracum is a very glossy light brown, and projects far beyond the shell margin. It is slit by lines radiating from the beak. The interior is light purplish.

The ragged, fringed appearance which the free edge of the periostracum gives the shell, justifies the popular name of awning shell. We collected this species at Cold Spring Harbor and at Ward Point near Tottenville on Staten Island. It can be found in bays along the muddy bottoms, where it flits through the water by opening and snapping shut its valves like a *Pecten*.

NUCULA PROXIMA Say *Nut Clam*

SIZE: Up to ½ inch in diameter.
DESCRIPTION: The shell is thick, solid, and obliquely triangular in outline. The lower edge is finely crenulated, and in life the entire shell is covered by an olive green periostracum. The interior of the valves are of pearly nacre. The teeth are comb-like as in *Arca*, but are arranged along two sides of an acute angle.

This shell lives in large colonies in the mud in moderately deep water where, with luck, large numbers can be dredged. We

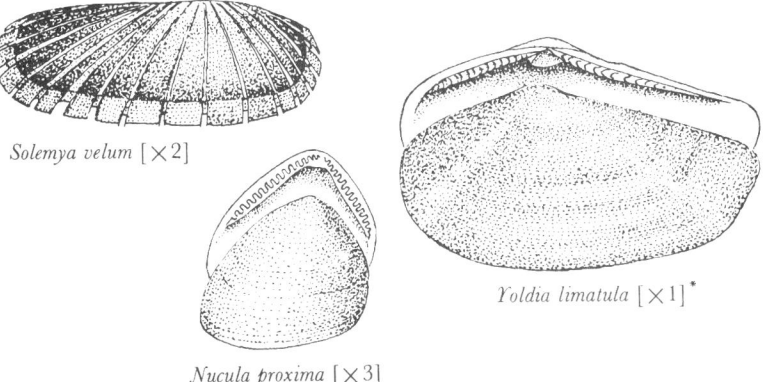

Solemya velum [×2]

Nucula proxima [×3]

Yoldia limatula [×1]*

* The figure of *Yoldia limatula* is not correct. The ventral margin is smoothly rounded from end to end; there is no sudden bend as shown in the illustration.

found such a colony near buoy 16 in Great South Bay, about one mile from Fire Island in about ten feet of water. Dead valves are infrequently found on the beaches.

YOLDIA LIMATULA Say *Shiny Nut Clam*

SIZE: 2 inches long, one inch wide.
DESCRIPTION: The shell is thin and covered with a highly glazed, light green periostracum, occasionally with lighter or darker zones or radiations. The front end and base are regularly rounded, the hind end drawn out into an obtuse point. The interior of the valves is bluish white and there is a centrally placed cartilage pit at the hinge line. The teeth are arranged like the teeth of a comb on both sides of the cartilage pit.

We once found a veritable sliver of a shell at Orchard Beach near Pelham, merely the hinge portion of one valve of this small clam. However, this was quite enough to permit us to recognize a *Yoldia*, the only "specimen" of this gleaming clam we have ever actually collected. It was said to be abundant many years ago in Raritan Bay and older collectors found it to be rare to moderately abundant near Long Island and Staten Island. It probably still lives there.

YOLDIA SAPOTILLA Gould is similar but smaller and lighter in color. It is about 1 inch long and ¾ inch wide. It has only 16 to 18 teeth on each side of the cartilage pit, whereas *limatula* has 22 on the anterior side, and 18 on the posterior.

Both species live on clean sand or in sandy mud in water deep enough to require dredging.

ANADARA PEXATA Say* *Blood Ark*

SIZE: 2¼ inches long, 1¼ inches high.
DESCRIPTION: The ark shells are heavy, deeply ridged shells with numerous tiny comb-like teeth arranged in a line along the hinge margin and appearing in both valves. When alive the shells are covered with a thick, dark brown periostracum, but this wears off rapidly in dead shells. The beaks are in very close contact and are sharply bent towards the front end of the clam. The shape is broadly

---

* See List of Changes in Nomenclature, page iii.

oblong with the upper and lower margins definitely rounded.

The ark clams are known popularly as "blood clams" because their blood, instead of being bluish in color, is a watery red. Single valves are common on all outer beaches, and fine live specimens can be collected at Shinnecock Bay and on the south shore of Staten Island. We have also dredged them from about 5 feet of water in Great South Bay and elsewhere. In the Spring we sometimes find tiny live specimens on the beach at Rockaway and Coney Island.

In a recently published shell book, the name of this shell appears as *Anadara ovalis* Bruguiere. Opinion regarding the validity of this change of name is divided. It used to be placed in the genus *Arca*.

ANADARA TRANSVERSA Say *Transverse Ark*

SIZE: 1½ inches long, 1 inch high.

DESCRIPTION: The shell is heavy, white, deeply ridged. The outline is definitely oblong, the upper and the lower margins being practically straight. The beaks are not in close contact, but are separated by a narrow space; they are only slightly or not at all directed forward.

We have collected this species together with *A. pexata*. Until recently this shell was called *Arca transversa* Say.

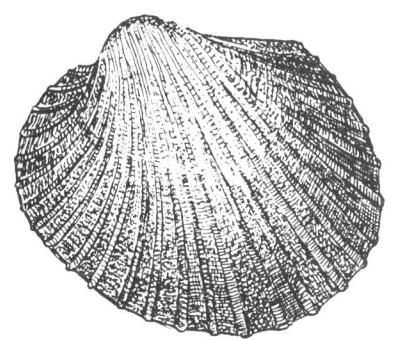

Anadara pexata [×1]

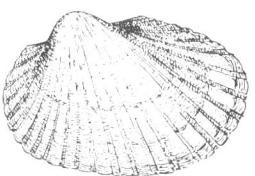

Anadara transversa [×1]

NOETIA PONDEROSA Say   *Ponderous Ark*

SIZE: 2½ inches long, 1½ inches high.
DESCRIPTION: The shell is deeply ridged, thick and heavy. The beaks are widely separated by a flat, strongly striated hinge area. There is a rounded ridge at the hind end of the shell which causes this end to descend almost perpendicularly. The lower edge is nearly straight with a gentle bend in the middle.

The last of our ark shells has not been found alive north of Cape Hatteras, but dead single valves are not uncommon on the south shore beaches of Long Island. Like *Littorina irrorata,* they probably died out here as the waters turned colder. In Florida, where live *N. ponderosa* are common, the shell is covered by a thick, blackish-brown periostracum.

CRASSOSTREA VIRGINICA Gmelin   *Oyster*

SIZE: Commonly 2 to 6 inches long, width various.
DESCRIPTION: The valves are rough, grayish-white outside, white in the interior. The muscle scar is usually purple.

The common oyster is the most valuable, best known and most thoroughly studied of all our mollusks. In the wild state it is found attached to all manner of objects, even to the live shells of relatively active periwinkles (*Littorina*). When allowed to grow unmolested it becomes a massive, elongated object as much as ten inches long with valves more than an inch thick. But the oyster has so many enemies, echinodermian, molluscan and human (who cultivate it only to devour it) that most end their lives long before they reach such a patriarchal state. The oysters we are best acquainted with in restaurants are about three years old. In our area we find oyster shells everywhere on beaches and even inland, where the aboriginal Indians left mounds of oyster shells, the remains of pre-Columbian feasts. Such remains can be seen in Inwood Park in upper Manhattan and at Ward Point in Staten Island. Several areas in the bays off the South Shore support flourishing oyster fishery industries. The story of oyster culture is an interesting one, and both the state and federal governments have published valuable pamphlets about it.

Until recently this species was known as *Ostrea virginica.*

PECTEN IRRADIANS Lamarck — *Bay Scallop*

SIZE: 2¼ inches in diameter.
DESCRIPTION: The common bay scallop is round with a flat hinge margin ending in unequal ears. It is "scalloped" by 17 to 20 wide, rounded and elevated ribs. The color is variable, usually a dusky or blackish slate, but orange, reddish, purplish or yellow-brown specimens are found. Frequently the lower valve is white, the upper darker, and some specimens have both valves white. The upper valve is slightly but noticeably more convex than the lower.

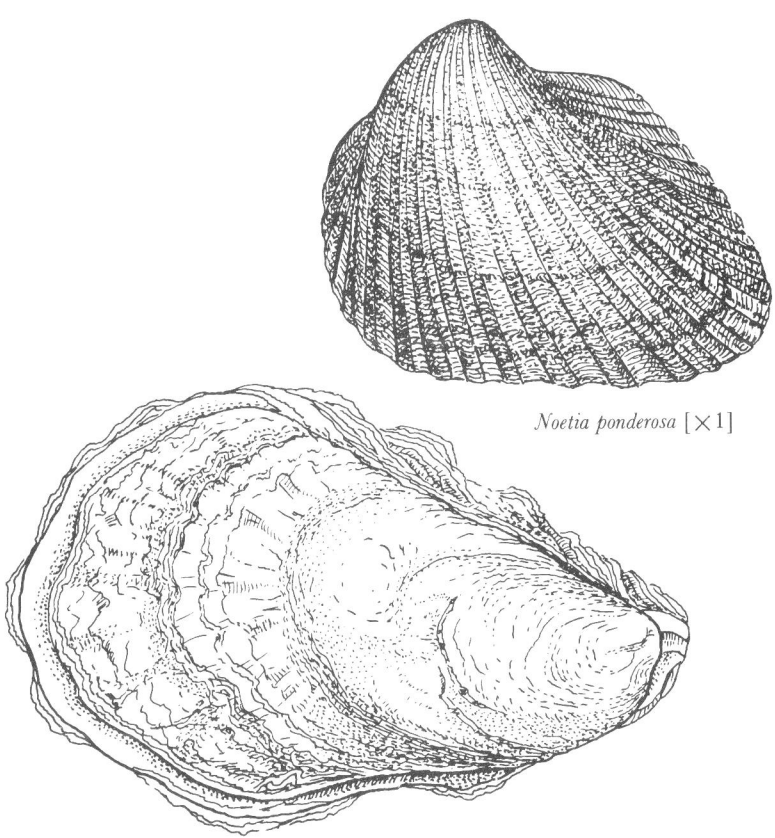

Noetia ponderosa [×1]

Crassostrea virginica [×½]

This scallop lives in bays among the eel grass. The mantle of the living animal is ornamented with a series of brilliant blue ocelli which serve as weak eyes, strong enough to distinguish between dark and light objects. The scallop can move about quite rapidly, if erratically, in the water by opening and shutting its valves, thus shooting off in any direction when danger threatens. It is extensively fished in season to provide a delectable sea food, but only the powerful muscle, called the "eye" by the trade, is eaten, the rest being discarded as the scallop fisherman sails toward shore. Dead, discolored single shells are quite common on all outer beaches. We found it alive in shallow water near the shore at Orient Point on Long Island. It was recently placed in the genus *Aequipecten,* but not all experts agree.

PECTEN MAGELLANICUS Gmelin   *Ocean Scallop*

SIZE: 5 to 6 inches in diameter.

DESCRIPTION: The upper valve is concave, light brown or flesh colored, and densely covered by numerous fine lines

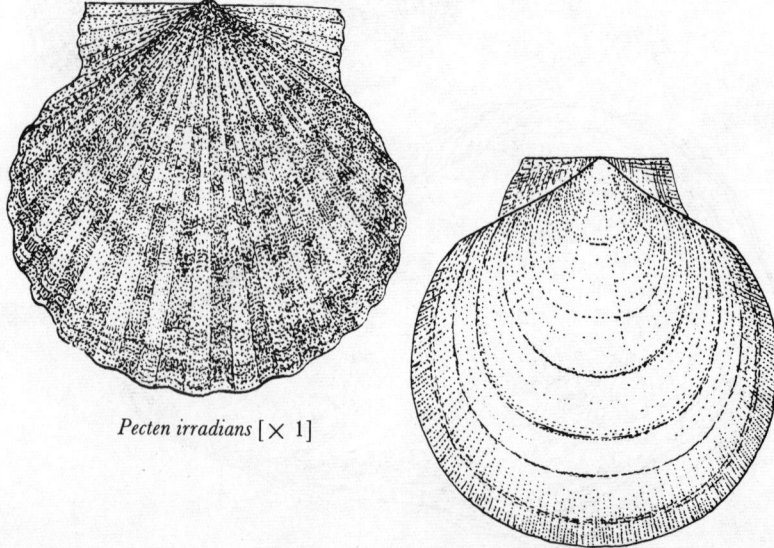

Pecten irradians [× 1]

Pecten magellanicus [× ½]

radiating from the beak. The lower valve is white and nearly flat, and shows the radiating lines much more weakly. In mature specimens the valves, especially the upper, are much chewed away and eroded as a result of the activities of *Cliona*, a boring sponge. The interior of both valves is white, smooth and shiny.

The young of this shell is delicate and translucent and when first discovered in the stomachs of fish caught off the Massachusetts coast was thought to be a different species. Single valves of this scallop, in more or less broken condition, are picked up at times at Rockaway and Coney Island. Once in the 1920's a strong winter storm stranded a large colony of them on the beach near Neponsit, providing a number of free scallop dinners for the local inhabitants.

This widely ranging ocean species was first described from a shell taken in South America, whence its specific name. For a time our species was called *P. grandis*, but it had to take the older name when it was found to be identical. It was recently placed in the genus *Placopecten*, some believe unnecessarily.

ANOMIA SIMPLEX d'Orbigny *Jingle Shell*

SIZE: 1 to 3 inches in diameter, but in our area frequently smaller. DESCRIPTION: The common jingle shell is almost too well known to need description. Single valves of various

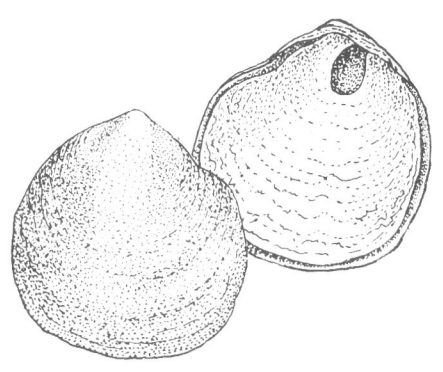

Anomia simplex [×1]

Anomia aculeata [×1]

colors and innumerable shapes are found on every beach and are usually among the first specimens a new collector picks up. The shell can be lemon-yellow, golden, brownish or pale buff, but always possesses a peculiar sheen. The great variation in shape is due to its habit of conforming to the surface upon which it lives. However, these gleaming shells are only the upper valves of the animal, the lower valve being flat, white, calcareous and possessing a large hole near one end through which the animal constructs a hard, solid peduncle that anchors it immovably to the object upon which it lives. These lower valves are much more rarely found on the beach.

Anomia attaches itself to any solid submerged object, preferring stones and dead or live shells. We have frequently taken live specimens from the shells of oysters on sale in fish stores. In Rockaway, brilliant black specimens are frequently found, but these are merely dead shells that have been trapped in black mud and have been thoroughly stained by it.

ANOMIA ACULEATA Gmelin is a much smaller, more delicate shell than *A. simplex,* and its yellowish-white surface is roughened by radiating, prickly scales instead of being smooth and shiny. *A. aculeata* reaches a length of about one-half inch. Apparently this species is more common north of the New York area, for we have collected only a few specimens on Long Island.

⁓ MYTILUS EDULIS Linné *Common Mussel*

SIZE: 1 to 3 inches long, ¾ to 1 inch at its widest point.
DESCRIPTION: The shell is violet-blue, elongate triangular, its pointed beaks at the upper extremity. The periostracum is blue-black and shining. A number of tiny teeth occur at the very tip in the interior of both valves.

The common mussel attaches itself by means of a strong thread-like anchor, called the byssus, to all sorts of objects that are laid bare by the retreating tide. Here it can be seen in immense numbers of all sizes and shapes, thickly covering jetties, rocks, bridge supports and other submerged structures. After storms large masses, securely held together by the tenacious byssus, are thrown on the shore. The species is extensively distributed throughout all northern seas and is much esteemed as a food in England, France and Russia. In the Spring, every bit of flotsam thrown ashore will be found to swarm with tiny, attached mussel shells.

In every large mass of *Mytilus edulis* one may find many shells of a lighter color, frequently beautifully decorated with bluish rays diverging from the beak. This form is called *Mytilus edulis pellucidus* Pennant, but whether it actually is entitled to a subspecific name is questionable. It appears to be a variety that, not being localized but rather found in every colony, should not be considered a valid subspecies.

We cannot easily forget a group of gypsies who came one Sunday to spend the day at Coney Island. Under their brilliant clothes they wore their bathing suits and each one brought along, wrapped in newspapers, a half loaf or so of bread and an onion. When lunch time arrived, they cut open the onion, rubbed it diligently onto the bread and then sauntered to the rock breakwaters which the retreating tide had uncovered. Here with a strong knife each one cut loose a dozen small mussels, cut open the tight shells and sucked in the yellow contents between big bites of onion-covered bread. They ate with such gusto and apparent delight that it took much willpower to keep from asking them for a snack. We still reserve the right sometime to indulge in such a meal!

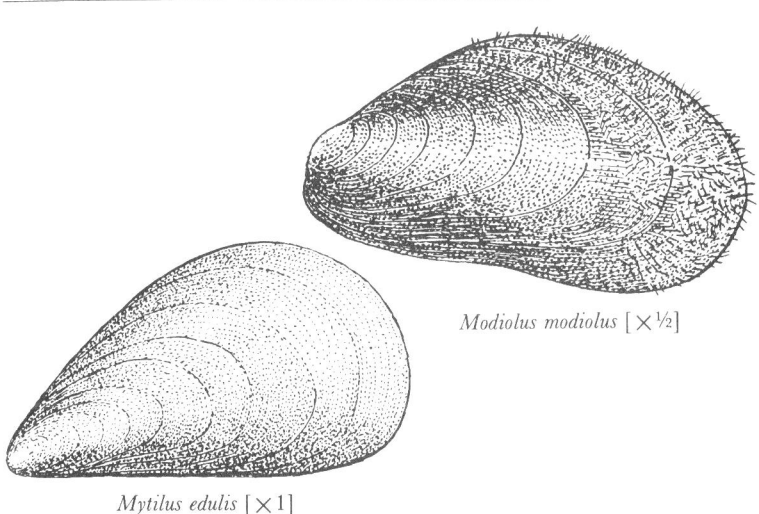

Modiolus modiolus [×½]

Mytilus edulis [×1]

MODIOLUS MODIOLUS Linné *Horse Mussel*

SIZE: 4½ to 6 inches long, 3 to 4 inches at the widest point.
DESCRIPTION: The horse mussel has a large, heavy, coarse shell covered by a leathery, glossy chestnut-brown periostracum. The beak is not at the very tip as in *Mytilus;* a rounded extension of the lower margin extends a little beyond it.

This mussel dwells in deep water, but is frequently thrown upon Long Island beaches by strong storms. It is far rarer than either *Mytilus* or *Brachidontes* and, unless they are dredged, fresh specimens are found only as a matter of very good luck. We have collected shells at Long Beach, Atlantic Beach and elsewhere on Long Island.

The genus *Volsella,* in which this and the following species were recently placed, has been found to be invalidly employed.

BRACHIDONTES DEMISSUS PLICATULUS Lamarck *Ribbed Mussel*

SIZE: 3 to 4 inches long, 2 to 2½ inches at the widest point.
DESCRIPTION: The ribbed mussel has crowded, raised, diverging ribs that thickly cover the wider end of the elongated triangular shell. Its beak is in the same position as in *Modiolus.* The periostracum is thin, very brilliant, yellow, green and brownish in color, usually arranged in broad, bent bands.

This species inhabits the rivulets of the salt marshes where it is found crowded in among stones or imbedded in the peat-like soil of the islands. "In this position, with the upper posterior portion slightly exposed, they crowd in such numbers as to form a complete stratum from six to twelve inches in thickness. A great portion of the time they are, of course, out of water; but they retain enough to serve the demands of their economy during the recess of the tide." (Gould). It may be found in vast numbers in Jamaica Bay and every similar marsh.

The name of this mussel seems at the moment to be in some state of confusion. For years it was called *Modiolus demissus plicatulus,* and the recent attempt to call it *Volsella demissa* Dillwyn has been generally rejected. The present generic assignment may still have to be changed.

CONGERIA LEUCOPHAEATA Conrad

Brackish-water Mussel

SIZE: ½ inch long, ¼ inch wide.

DESCRIPTION: The shell looks like a small *Mytilus*. It is inflated, elongate triangular, bluish white to tan in color, covered by a grayish brown periostracum that is roughened by concentric, slightly raised ridges. There is a white septum in the interior near the pointed beaks.

This shell was long thought to range no further north than Chesapeake Bay. However, we found a huge colony at Croton Point State Park, on the Hudson River, living attached to submerged buoys, logs and diving floats. We also found specimens all along both banks of the Hudson, from just south of the Bear Mountain Bridge down to Dyckman Street in upper Manhattan. There are no other published records of this species from our area.*

PANDORA GOULDIANA Dall

Gould's Pandora

SIZE: 1½ inches long, little more than ¾ inch high.

DESCRIPTION: The valves are pearly under a thin, chalky surface layer, very compressed, rounded at one end and

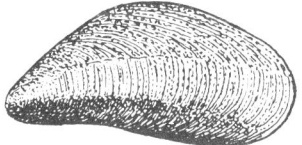

Congeria leucophaeata [×3]

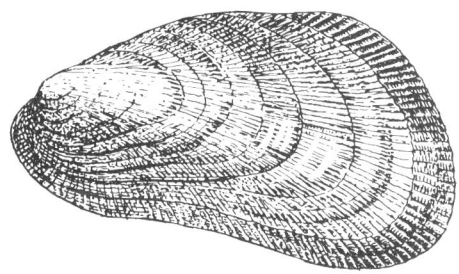

Brachidontes demissus plicatulus [×⅔]

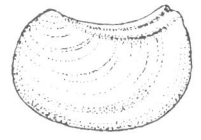

Pandora gouldiana [×1]

* *Congeria leucophaeata* was reported at Haverstraw, New York, in 1953.

tapering to an upward pointing peak at the other. The right valve is quite flat, the left somewhat convex and the hinge margin strongly flattened and sharply set off from the rest of the shell by a sharply cornered ridge.

The most surprising feature of this clam is its completely compressed form which makes it look like part of a broken single valve rather than a complete shell. We have dredged it in shallow water in Cold Spring Harbor and other similar places on the Sound. Living specimens recently were dredged in 10 to 15 feet of water in Peconic Bay, just off the high bluffs at Hampton Bays. Dead shells are frequent on the beach at Rockaway, but it takes a well-tutored eye to recognize these irregular fragments as parts of *Pandora* shells. The easiest indication of this is the upper, deeply curved margin which, since it is bordered by a heavy ridge of shell matter, keeps its shape longer and lets one know that *Pandora* is involved.

PERIPLOMA LEANUM Conrad *Lea's Spoon Clam*

SIZE: 1 inch long, ½ inch high.
DESCRIPTION: The shell is oval, thin and brittle; white with a thin yellowish periostracum. A spoon-shaped projection extends downward from the umbo in each valve.

This frail little shell is sometimes found thrown up by the pounding billows at Hither Hills Park and elsewhere on the outer beaches of Long Island. One is amazed to see how it can survive more or less intact the tremendous force of the waves which cast it up from about two fathoms, where it lives.

LYONSIA HYALINA Conrad *Glassy Clam*

SIZE: ¾ inch long, about ¼ inch high.
DESCRIPTION: The shell is elongated, the beaks prominent and located nearer the rounded anterior end; the posterior portion is drawn out into a narrower section, squarely cut at the end. The valves are thin, glassy, pearly, inflated; the periostracum is drawn up into fine radiating wrinkles which are most prominent along the lower portion. The periostracum has minute fringes that trap grains of sand to cover the shell.

The animal lives in shallow water, imbedded upright in firm bottom clay. We once collected about 200 specimens in Oyster Bay. We brought up large sections of the bottom with a clam hook and then proceeded to pick these mollusks at our leisure. We have also strained specimens in shallow water at Cold Spring Harbor and in Great South Bay. The thin, glassy shell structure is so characteristic that even fragments can be identified as belonging to this species.

ARCTICA ISLANDICA Linné   *Black Quahog*

SIZE: 3½ inches in diameter.
DESCRIPTION: The shell is almost circular in outline, thick, heavy, chalky. The color is white and a strong, black-brown periostracum covers the entire shell. The presence of this periostracum distinguishes *Arctica* from *Venus,* which it superficially resembles. *Arctica* also differs in the nature of its hinge teeth, the lack of any purple shell coloring, and the more rounded outline.

The black quahog is abundant in moderately deep water in New England. It is quite rare south of Massachusetts, but we

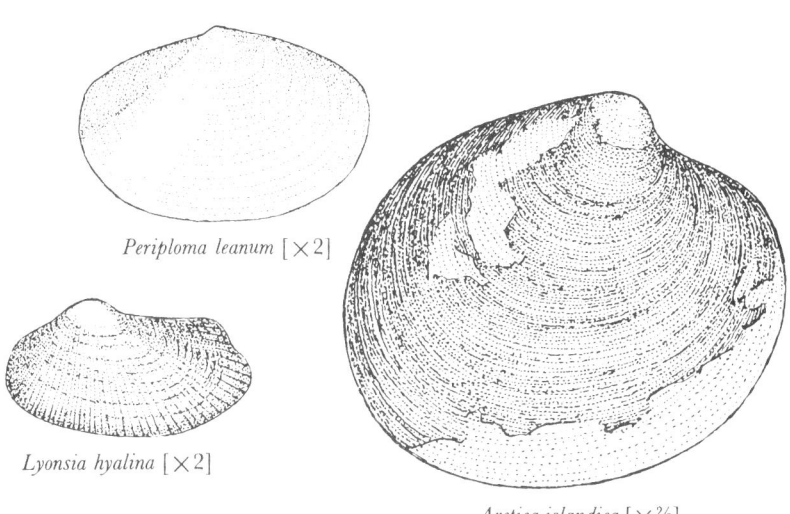

Periploma leanum [×2]

Lyonsia hyalina [×2]

Arctica islandica [×⅔]

have collected several dead valves on the beach at Hither Hills State Park and even two small valves in Rockaway. In more northern waters, this clam is harvested and serves as a moderately important source of sea food.

ASTARTE CASTANEA Say — *Smooth Astarte*

SIZE: 1 inch in diameter.

DESCRIPTION: The valves are heavy, thick, covered with a smooth, moderately glossy, light brown to black periostracum. The beaks are high and slightly hooked. The inner margins are finely crenulated.

Dead, worn shells are common at Rockaway, and we have frequently collected fine live specimens after winter storms at Long Beach and especially at Point Lookout before the rock jetty was constructed. Particularly fine specimens are said to live at Sandy Hook. In life the animal is quite active, moving along vigorously by means of its brilliant vermilion-colored foot. Shells having a tarry black periostracum have been named *Astarte castanea picea* by Dr. Gould.

Much less frequently we have found shells that resemble this *Astarte*, but whose surface is sculptured with rounded, concentric ridges separated by deep grooves. This is the species described by Dr. Gould as ASTARTE UNDATA.

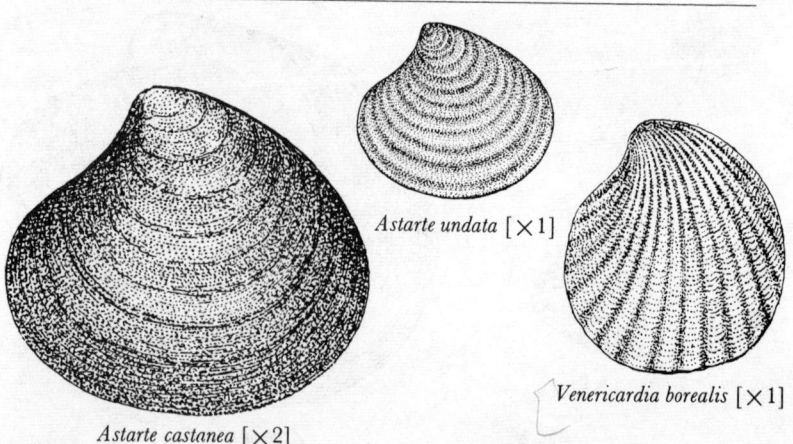

Astarte undata [×1]

Venericardia borealis [×1]

Astarte castanea [×2]

VENERICARDIA BOREALIS Conrad   *Cod Clam*

SIZE: 1 to 1½ inches in height, 1 inch wide.
DESCRIPTION: The cod clam has a heavy, solid shell, covered, even in beach-worn specimens, with a thick, persistent, rusty brown or black periostracum. There are 18 to 20 rather high, rounded radiating ribs with narrow spaces between. The margin is crenulated by ribs, the interior is a dull white in color.

This dweller of deep off-shore waters is most easily collected from the stomachs of cod fish, for which it apparently forms a steady article of diet. In Rockaway and on the outer Long Island beaches single valves are frequently cast up. The radiating ribs make this shell distinctive enough to keep it from being confused with any other in our area.

Cross-hatched Lucina
DIVARICELLA QUADRISULCATA d'Orbigny

SIZE: About 1 inch in diameter.
DESCRIPTION: The shell is white, glossy, swollen, nearly circular but somewhat flattened along the hinge margin. It is sculptured with strong concentric lines crossed by regularly spaced, nearly parallel grooves that meet in a wide angle

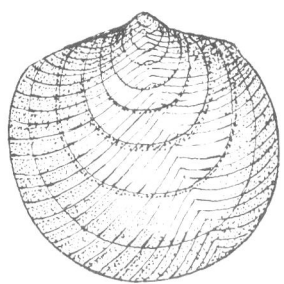

Divaricella quadrisulcata [×2]

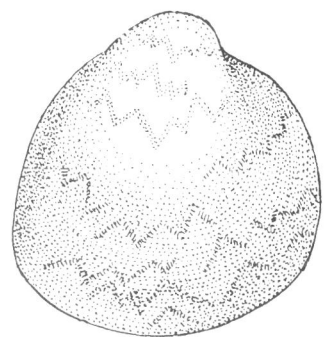

Laevicardium mortoni [×2]

along the front third of the shell and from this point are bent obliquely downward toward both ends of the valves. There is some resemblance to closely spaced military chevrons.

The hinge is very weak, for though we have collected many scores of valves, we have never come across a joined pair. It lives in deeper off-shore waters and enjoys a very wide distribution along the entire Atlantic coast. The circular outline, glossy texture and peculiar sculpture eliminate any chance of error in recognizing this species.

LAEVICARDIUM MORTONI Conrad   *Duck Clam*

SIZE: 1 inch high, about ¾ inch wide.
DESCRIPTION: The inflated shell is obliquely triangular, the lower margin rounded, thin, faun colored or dirty white with a very thin, rather glossy periostracum. Sometimes there are zigzag markings of brown on the surface. The interior is pale yellow with a purplish spot on one margin.

This is the only true *Cardium* or cockle shell in our area. It is very common in shallow water in Peconic Bay, Cold Spring Harbor and Great South Bay. We collected large numbers on Shell Beach and near Owl's Head on Shelter Island. Even the beach at Rockaway yields a few single, discolored valves. On Long Island it is said to form part of the diet of ducks, from which it derives its local common name of duck clam.

VENUS MERCENARIA Linné   *Hard-shell Clam*

SIZE: 3 to 5 inches in length, 2 to 4 inches wide.
DESCRIPTION: The solid, heart-shaped shell has prominent beaks; the color is grayish, the interior white with a large purplish area. The inner edge is finely crenulated.

The common hard-shell, little neck or cherrystone is too well known to require describing. The young have very prominent, widely spaced, concentric blade-like ridges. It lives in the mud bottom in shallow water in all bays along Long Island, where it is fished for by means of large, rake-like clam tongs or trawl nets. Its Latin name was given to it by the father of scientific nomenclature, Carl Linné, when he learned that the shell was ground into cylindrical beads from which the Indians made their wampum or money.

Sometimes a clam is found that is beautifully marked with zigzag lines of light brown. This form is called VENUS MERCENARIA NOTATA Say. Shells that lack the purple spot were named *V. mercenaria alba* Dall.

This species was recently placed in the genus *Mercenaria*.

PITAR MORRHUANA Linsley *Thin Venus Clam*

SIZE: 2 inches long, 1½ inches wide.

DESCRIPTION: The shell is comparatively thin, moderately inflated, and roughly resembles the hard-shell clam. The color is dirty white with several rust-colored or dark gray

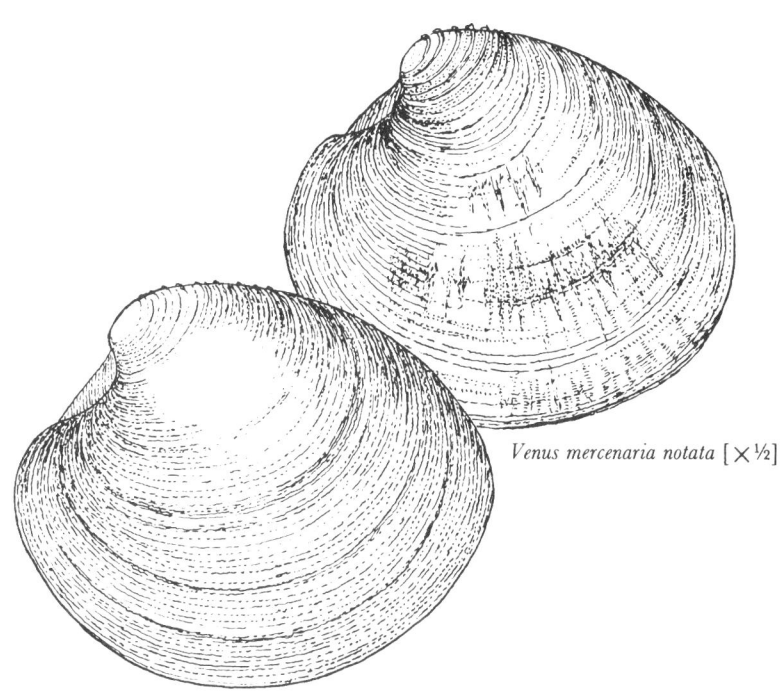

Venus mercenaria notata [× ½]

Venus mercenaria [× ½]

areas. The margins are smooth, not crenulated, and the interior never has a purple spot.

This species is sometimes mistaken for *Venus,* but the thin shell, the lack of purple color and the smooth margins readily set it off. Perfect specimens are occasionally found on the beach at Rockaway and elsewhere on Long Island. It lives in shallow, off-shore waters.

GEMMA GEMMA Totten *Gem Clam*

SIZE: Very small, ⅛ inch in diameter.

DESCRIPTION: The valves are trigonal, moderately thick with a shiny surface crowded with irregularly spaced, concentric furrows. The color is whitish to tan with a purple flush over the beak and posterior areas. The lower margin is crenulated.

This tiny relative of *Venus*—it was long thought to be nothing more than its young—is very common in most bays, living in sand from just below the high water mark to the deeper waters off-shore. We have it from some 20 different localities in Long Island and New Jersey as well as within New York City limits. Specimens can be found among the debris left by the high tide on beaches, or can easily be dredged in shallow water with a strainer tied to a stick.

The subspecies GEMMA GEMMA MANHATTENSIS Prime, was found originally in 1852 in the East River near Hell Gate, where we are sure no mollusks of any kind have lived for generations. It is more triangular in outline; the surface furrows are deeper and more regularly spaced; and the purple color is completely absent. We have found shells that seem to belong to this form at Howard Beach, near the causeway that carries Cross Island Boulevard to Rockaway.

GEMMA PURPUREA Lea is a similar shell, but less inflated, less triangular in shape and with less purple in the shell than *G. gemma*. We found specimens in Sunken Meadow State Park in Long Island. Many malacologists believe that these species are not valid, and represent merely forms of *Gemma gemma* Totten.

General Joseph G. Totten, the author of *Gemma gemma* and other local shells, and in whose honor Tottenville, Staten Island, and Fort Totten are named, lived a most interesting life. He was born in New Haven, Connecticut, in 1788 and died in Washington in 1866. All his life he was attached to the military forces and was instrumental in strengthening our coast defenses when the invention of steam-driven warships posed a new threat. He wrote extensively on the engineering problems involved in casement and light-house construction, personally conducting experiments to test the relative strength and utility of various kinds of stone and wood. In 1855 he was chosen by the New York State legislature to preserve "the harbor of New York from encroachments and to prevent obstruction to the necessary navigation thereof." Here he improved navigation around Hell Gate and helped preserve the important Gowanus Bay.

In addition to his engineering projects he was interested in mineralogy and structural geology. As a young man he described the prehistoric Indian mounds which he explored while on a surveying trip to Ohio. He also discovered and named some nine or ten species of local shells himself, but found many more which he submitted to Dr. Gould who then described and named them. He served as Regent for the Smithsonian Institution from 1846 on, and at his suggestion, the new building was provided with fire-proofing to protect the more valuable exhibits. Upon his death, the Smithsonian received his valuable collection of shells and minerals.*

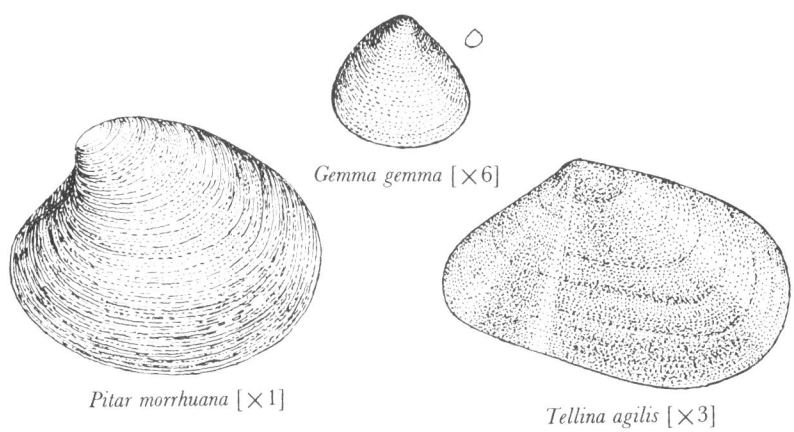

Pitar morrhuana [×1]

Gemma gemma [×6]

Tellina agilis [×3]

* A related species, *Mysella planulata*, is described in the Supplement of Additional Species, page 137.

TELLINA AGILIS Stimpson — *Fragile Wedge Clam*

SIZE: A little more than ½ inch long, ¼ inch wide.
DESCRIPTION: The valves are moderately elongated, compressed, very thin, fragile, translucent, white or with more or less large areas of pink. The surface is faintly iridescent, shiny, covered with very fine concentric impressed lines.

This species is found the year round in Rockaway in moderate numbers, in wet sand. Most of these specimens are beautifully tinged with pink. In the bays of Long Island the shell is predominantly white, sometimes faintly yellow. The animal is a shining white and possesses delicate and very long siphons which extend up through the sand to the surface in order to take in fresh oxygen and food-bearing water and to eject waste products. Until recently it was called *T. tenera* Say.*

MACOMA BALTHICA Linné — *Baltic Macoma*

SIZE: 1 inch long, ¾ inch wide, rarely larger.
DESCRIPTION: The valves are compressed, broadly oval in outline, but somewhat narrowed and pointed at one end. The beaks, almost centrally placed, are tiny. The color is usually dull white, but the specimens collected at Orchard Beach near Pelham have a delicate rose tinge in the rather chalky shell. There is a gray or faintly yellow periostracum, usually worn off near the beaks.

This species lives in the quiet water of the bays, but near enough to the shore to be thrown on the beach in large numbers. We have it from many bays on both shores of Long Island, and from Ward Point in Staten Island.

As can be seen from its name, the clam is also common in the Baltic Sea in Europe. The modern spelling does not include the "h", but it is common practice in nomenclature to preserve the name given by the original author, errors and all.

MACOMA CALCAREA Gmelin is larger and more elongate than *M. balthica* (1½ to 2 inches long). It is chalky, white, with a dingy periostracum. It is best distinguished by the upper margin of the pallial sinus, which in this species runs more or less parallel to the lower shell margin, whereas in *M. balthica* it descends rapidly. This species is fairly common in the muddy bottoms in shallow water in Cold Spring Harbor.†

---

* A related species, *Cumingia tellinoides*, is described in the Supplement.

† Another *Macoma* species is described in the Supplement of Additional Species, page 137.

DONAX FOSSOR Say*   *Wedge Clam*

SIZE: About ½ inch long, ¼ inch wide.

DESCRIPTION: The valves are rather thick, the beak placed at one end giving the shell a wedge-shaped outline. The surface is shiny, with finely incised radiating lines which are frequently completely covered by a thin, shining layer. The color is quite variable, with rays or blotches of purple, orange-yellow or blue on a white background, though entirely white shells are not infrequent. This color variation is much more marked in the interior of the valves. Color pigments tend to fade in dead shells.

In some years in late summer, we have found thousands of these shells in Rockaway uncovered by the breaking waves. However, as soon as a wave retreats for its next assault, these small jewel-like clams stick out a tiny, agile foot, up-end themselves and in a flash disappear once more into the wet sand. Unfortunately no specimens have appeared on the beaches in our area since 1955, but undoubtedly they will return soon. They always have in the past. Because of this strange periodicity, some malacologists speculate that what we call *D. fossor* is merely *D. variabilis* (a large, Floridian *Donax*), the larvae of which have somehow been swept far to the north and developed poorly in our unfavorable waters. It is fairly certain that *D. fossor* does not survive the winter in our area.

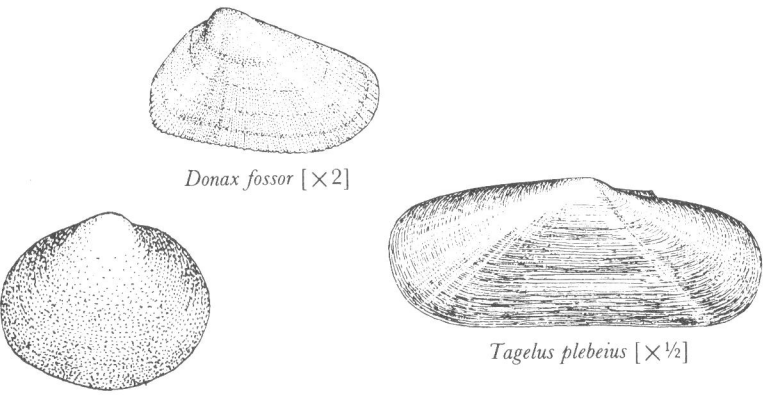

Donax fossor [×2]

Tagelus plebeius [×½]

Macoma balthica [×1]

---

* Biological data indicate that populations of *"Donax fossor"* in this area represent a northern, depauperate growth form of *Donax variabilis* Say.

TAGELUS PLEBEIUS Lightfoot   *Blunt Razor Clam*

SIZE: 4 inches long, about 1 inch wide.
DESCRIPTION: The valves are narrow, oblong, moderately inflated. The upper and lower margins are nearly parallel, the two ends rounded. The low beaks are just off center. The color is chalky white and there is a thin, yellowish periostracum.

This species lives well buried in sand or mud in the tidal zone of the bay beaches in shallow water. In Rockaway we frequently find single discolored valves. Until recently this species was called *T. gibbus* Spengler.

TAGELUS DIVISUS Spengler is a smaller shell (1½ by ¾ inch) with bowed instead of straight upper and lower margins and a thin, straw-colored periostracum. It is reported from several areas in eastern Long Island.

SILIQUA COSTATA Say   *Fragile Razor Clam*

SIZE: 2 inches long, ¾ inch wide.
DESCRIPTION: The valves are elliptical in outline, thin, compressed, fragile, faintly translucent and covered with a shiny iridescent violet periostracum. The beaks are located off center, toward the front. In the interior of each valve there is a white, rather thick rib extending from the beak toward the lower margin.

The internal rib of this species is so characteristic, that even fragments of valves can be identified from it. The rib is narrowest and highest near the beak, from which point it widens and thins out till it disappears near the lower edge. It is so much stronger in relation to the rest of the shell, that fragments bearing it are not uncommon on Rockaway and other open beaches.

ENSIS DIRECTUS Conrad   *Common Razor Clam*

SIZE: Up to 10 inches long, 1¼ inches wide.
DESCRIPTION: The valves are very elongate, curving, squared off and gaping at both ends. They are covered by a brilliant, yellowish-green periostracum with a large,

elongated triangular purplish region near the curving edge. The shell looks very much like the handle of an old-fashioned straight razor, whence its popular name.

This clam is very active; it burrows with amazing speed downward into wet sand, and when placed horizontally into water, it leaps forward by a kind of water-jet propulsion. The arrow-shaped foot projects beyond the shell. Dead shells are frequently found on the ocean beaches, but the animal lives in sand and mud in the shallow water of the bays. We have found large colonies in Jamaica Bay, in Reynolds Channel at Long Beach, and in many other bays of Long Island.

Since the clam is one of our tastiest mollusks, it is frequently hunted for food. However, it has to be stalked with caution. Our method is not without success. When the razor clam burrows in the sand uncovered by the retreating tide, it leaves a small, keyhole shaped opening on the surface. One must approach this spot gently, count slowly to five (or to be safer, ten) and then quickly insert the tines of a pitchfork near the opening and throw a large clump of wet sand upward. Usually the mollusk, who has come near the surface when he thought the coast was clear—after having been warned of danger by footsteps—is thrown up with the first load. It is usually useless to try to uncover the clam again if the first attempt has not been successful. Likewise, it is quite futile to try to pull the animal from its burrow by hand. It holds on with such force that it is torn in two or the shell is broken before it can be removed.*

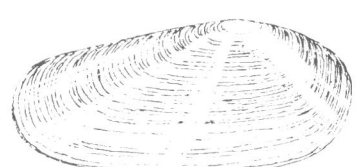

Siliqua costata [×1]

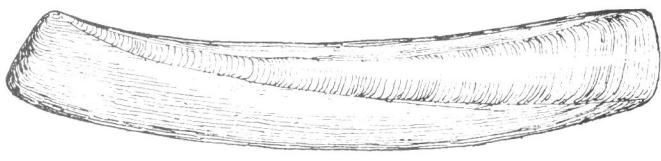

Ensis directus [×½]

* A related species, *Solen viridis*, is described in the Supplement of Additional Species, page 137.

SPISULA SOLIDISSIMA Dillwyn   *Surf Clam*

SIZE: Very large, up to 8 inches long, 6 inches wide. Most specimens are smaller. DESCRIPTION: The valves are somewhat triangular, massive, rather chalky, dead white in color, but frequently stained or partly covered by a dull, brownish or yellowish periostracum, most usually eroded away in beach shells. The inside, just under the beaks, has a large pit-like saddle in which rests a dark brown, resilient cartilaginous mass (resilium) that serves to cause the valves to spring open when the muscles relax.

This is the largest bivalve on the east coast. It is the most common beach shell in Rockaway and other ocean beaches. Winter storms throw it on shore in astounding profusion. It lives in the sand, usually below the low-water mark, and burrows slowly with its strong, ivory colored, tongue-like foot. It serves as an important article of diet for the local hordes of sea gulls who have learned the trick of picking up the clams in their claws, flying high with them and then dropping them upon the hard earth where they crack open and permit access to the soft food within. Sometimes the concrete bridge that leads from Far Rockaway to Atlantic Beach is literally covered with the remains of *Spisula* shells broken in this fashion. The young of this species makes a very tasty chowder.

MULINIA LATERALIS Say   *Small Surf Clam*

SIZE: ¾ inch long, ½ inch wide.
DESCRIPTION: The valves are small, thin, triangular, well inflated, white and moderately shiny. The surface is covered with a thin yellowish periostracum. The pit near the beaks which accommodates the resilient brown cartilage is quite deep. The posterior edge is somewhat flattened, set off from the rest of the surface of the shell by a rounded, prominent ridge that runs diagonally from the beak to the lower margin.

This small relative of *Spisula* lives in huge numbers in the muddy bottoms of bays. We have strained it in shallow water in Oyster Bay, Cold Spring Harbor, and Great South Bay. It is very common at Ward Point near Tottenville, Staten Island. The shell could be mistaken for the young of *Spisula,* but it is easily identified by its more swollen shape, deeper cartilage pit and flattened posterior edge.

Arctic Wedge Clam
MESODESMA ARCTATUM Conrad

SIZE: In our area 1½ inches long and 1 inch wide.
DESCRIPTION: The valves are heavy, compressed, white, chalky, and covered with a thin, light yellow periostracum. The beaks are placed far from the center, making the outline wedge-shaped. At the hinge there is a small, spoon-shaped cartilage pit as in the case of *Spisula,* to which the present shell is allied.

This rather large shell, that resembles *Donax* in shape, is common on the eastern beaches of Long Island. We have it from Hither Hills, Montauk Point, and as far west as Patchogue. But that seems to be as near as it ever gets to the New York city limits.

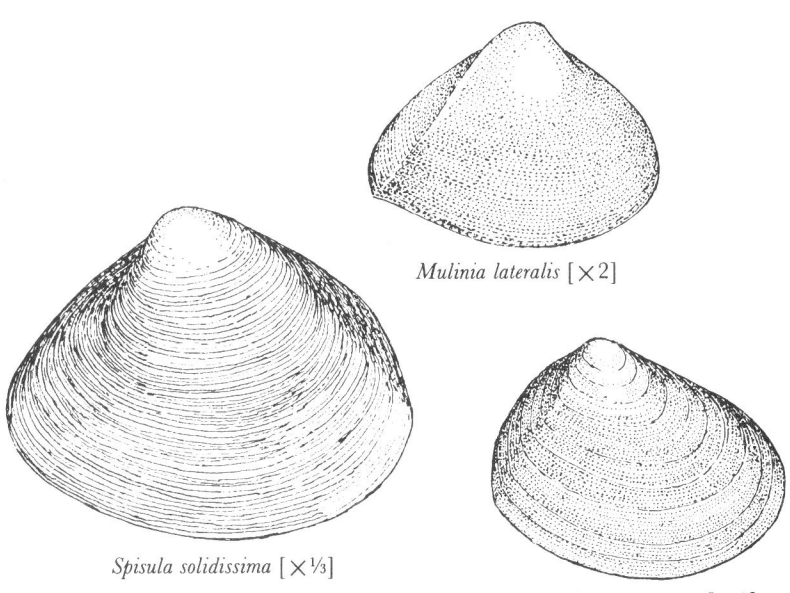

Mulinia lateralis [×2]

Spisula solidissima [×⅓]

Mesodesma arctatum [×1]

HIATELLA ARCTICA Linné   *Arctic Rough Clam*

SIZE: 1 inch long, ⅜ to ½ inch wide.
DESCRIPTION: The valves are coarse, chalky, white with a grayish periostracum. They are usually irregular, twisted and misshapen. The beaks are near the anterior end. The young are flatly oval, the upper and lower margins almost parallel, the posterior margin bearing a few short spines.

The shell has a very wide distribution over the entire northern hemisphere, but in the New York area we have succeeded in finding only a few valves on the beach at Far Rockaway. It can sometimes be taken on holdfasts of algae living offshore. It has been reported from Little Gull Island near Long Island. In older shell books it appears under the name *Saxicava arctica*.

CORBULA CONTRACTA Say   *Contracted Box Clam*

SIZE: ½ inch long, ¼ inch wide.
DESCRIPTION: The valves are inflated, moderately thick, white, beautifully sculptured by many regularly spaced, well impressed, narrow concentric lines. They are rounded in front and somewhat pointed behind. An angular ridge runs diagonally from the beaks to this posterior point. The lower edge is nearly straight, slightly raised in the middle. The left valve fits snugly into the right, instead of meeting it edge to edge as in most clams.

This species can be dredged in shallow water in the bays and in Long Island Sound. We have live-taken specimens from Shin-

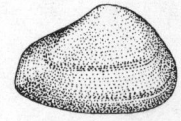

Corbula contracta [×2]

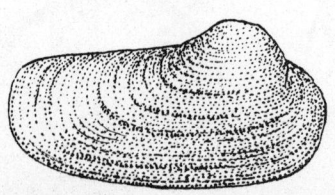

Hiatella arctica [×2]

necock Bay and Cold Spring Harbor, and single valves from the beach at Rockaway.

MYA ARENARIA Linné *Soft-shell Clam*

SIZE: In our area usually 3½ inches long, 2 inches wide.
DESCRIPTION: The valves are thin, wrinkled, widely elliptical, chalky, covered by a nondescript grayish periostracum. The valves gape at the ends. There is a long, spoon-shaped structure in the left valve in which the resilium rests. In life there are two long, joined siphons that extend from the shell and cannot be fully retracted.

This is the "steamer" or "soft shell clam" of New York, the "long clam" of eastern Long Island and the "clam" of New England, where the *Venus* clam is called a "quahog." It lives in a deep burrow in the mud, sand or gravel of bay shores and makes known its presence to the digger by shooting out jets of water from the siphons when it is disturbed by his footsteps. This habit gives it still another, somewhat less delicate, popular name. It is the most prolific mollusk in our area. Like the scallop, the oyster and the *Venus* clam, it is important enough to have several pamphlets on its life history printed by the Bureau of Fisheries in Washington and several state capitals.

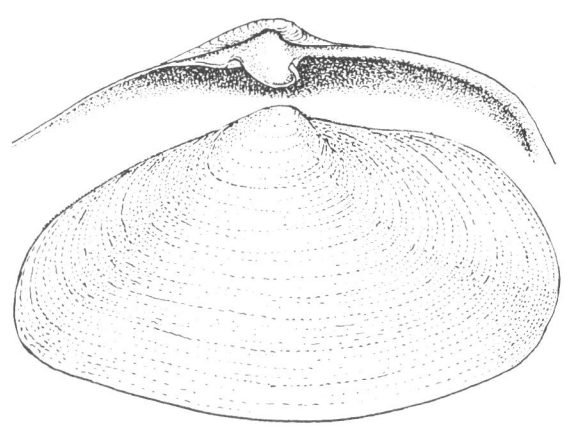

Mya arenaria [×1]

CYRTOPLEURA COSTATA Linné   *Angel Wing*

SIZE: 4 to 8 inches long, 2 to 3 inches wide.

DESCRIPTION: The large, thin white valves are marked by broad radiating ribs which are raised into tooth-like scales where they cross the rough growth lines. The interior is strongly marked with indentations that correspond to the exterior sculpture. The beaks are covered by a strongly reflected portion of the upper shell margin. Each valve possesses a narrow, spoon-shaped structure extending inward from the beaks.

The beautiful angel wing shell is usually found on Long Island in the form of fragments, which can be easily identified by the characteristic scaly ribs. Recently a live specimen was collected near Orient Point. In Florida, where the species is quite common, it lives burrowed several feet below the surface in loose sand.

This species was long known as *Barnea costata*.

PETRICOLA PHOLADIFORMIS Lamarck   *False Angel Wing*

SIZE: 2 inches long, 1 inch wide.

DESCRIPTION: The valves are chalky white, elongated, the anterior portion short and rounded, the posterior long and only slightly narrowed. The upper and lower margins are nearly parallel. The surface of the shell is ornamented with a series of lines radiating from the beak. Those on the

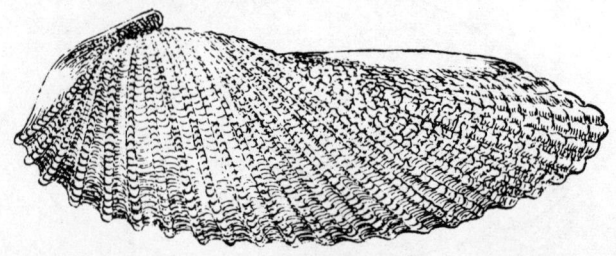

Cyrtopleura costata [× ½]

anterior end are much elevated and a row of raised, toothlike scales occurs where the concentric lines of growth cross the ribs. It is with these brittle but exceedingly hard scales that *Petricola* bores into the blue clay, waterlogged wood, or even rock in which it lives. The shiny white interior is marked with indentations that correspond to the exterior ribs.

Many specimens can be collected at Far Rockaway, especially on the narrow beaches that face the channel between Far Rockaway and Atlantic Beach. It is fairly common everywhere on bay and ocean beaches. There is a large colony living in the peat at Oakwood Beach on Staten Island.

Sometimes *Petricolas* are found that are shorter and broader in outline, with more numerous radiating lines and the hinge teeth shorter and grooved. These shells belong to the "subspecies" *P. pholadiformis lata* Dall.

BARNEA TRUNCATA Say *Small Angel Wing*

SIZE: About 2 inches long, 1 inch wide.
DESCRIPTION: The valves are white, the surface coarse with lines of small ribs that are also visible in the white, shining interior. The truncated posterior end is lightly covered by a thin, rust-colored periostracum. The anterior portion is pointed and the basal margin concavely

Barnea truncata [× 1]

Petricola pholadiformis [× 1]

arched. There is an internal, free rib-like projection extending from under the beaks. A portion of the upper margin is narrowly reflected over the beaks. A small, extra, arrow-shaped valve protecting the area between the two beaks can be seen in live specimens.

This small relative of the angel wing can be collected alive in banks of blue clay on the bay side just behind Lido Beach near Atlantic Beach. Dead valves are very common at Far Rockaway, which leads one to suspect the presence of colonies below the low water mark. We have frequently taken specimens from waterlogged wood thrown up on the beach at Rockaway. The animal is much larger than the shell and extends far beyond it.

ZIRFAEA CRISPATA Linné *Common Piddock*

SIZE: Up to 2 inches long, 1 inch wide.
DESCRIPTION: This borer superficially resembles *Barnea truncata* but can immediately be distinguished by the presence of a deep groove or furrow that runs obliquely from the beak to the middle of the basal margin and divides the valves into two almost equal portions. The area before this furrow is covered by coarsened, radiating ribs, the section behind is relatively smooth. The posterior margin is much more rounded than in *truncata*.

We have found this shell on the beach at Far Rockaway and have frequently taken specimens from burrows in waterlogged driftwood and floating masses of peat.

TEREDO species *Shipworms*

SIZE: The valves about ¼ inch in diameter, the tube various. DESCRIPTION: The clam has a long, wormlike body that is protected by a calcareous tube with which the animal lines its burrows. The front end is covered by two saddle-shaped valves that, under a magnifying glass, show the tiny chisels and rasps, beautifully arranged in serried lines, with which *Teredo* does its destructive work. The hind end, which is in contact with water, has two small siphons that can be withdrawn and protected by a pair of small, paddle-shaped pallets.

This is the notorious shipworm that causes so much damage to untreated wood pilings in all our salt water harbors. Many drift logs are found honeycombed with the burrows of this animal. Pilings so damaged are weakened until they collapse. The most common species in this area is *Teredo navalis* Linné.

CHAETOPLEURA APICULATA Say* *Pill Bug Chiton*

SIZE: 1 inch long, ¾ inch wide.

DESCRIPTION: The shell is flexible, composed of eight overlapping sections or valves held in place by a narrow, tough girdle ringing the outer margin. The valves are thickly sown with small bead-like points. The color is grayish or pale chestnut.

This is apparently the only representative of the Chitons, or coat-of-mail shells, living in our area. It inhabits shallow water affixed to pebbles or stones to which it clings vigorously if disturbed. When removed from its holdfast, it immediately begins to curl up and if it is allowed to dry in this condition, it cannot be straightened out again without breaking the valves. Sometimes the valves of dead animals are broken apart by wave action and

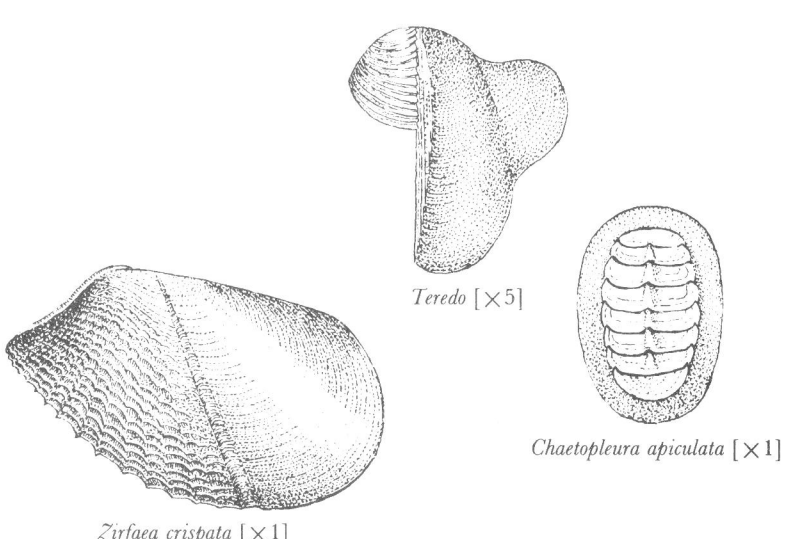

Teredo [×5]

Chaetopleura apiculata [×1]

Zirfaea crispata [×1]

* Strictly speaking, *Chaetopleura apiculata* should be in a separate subdivision of this chapter. It is not a bivalve, but belongs to the class of chitons, Polyplacophora.

when found in this condition are known as "butterfly shells" or "mermaid teeth."

It is said to be abundant along the coast of New York and New Jersey. Many were collected off Gardiners Island in Long Island Sound.

Collecting Shells in Fish Stores

Shell collectors in New York City soon find that live mollusks can be "collected" easily and cheaply in the many Italian fish stores that are found along Ninth Avenue between 37th and 42nd Streets and on First Avenue below 14th Street. We once made a census of all the live mollusks that can be found here and listed no less than forty-five species, including octopuses and squid. Though the most common species actually offered for sale are *Crassostrea virginica, Venus mercenaria, Mytilus edulis* and *Littorina littorea,* many small hitchhikers attach themselves, especially to the oysters. A careful scrutiny of the shaggy shells of these bivalves, with the indulgence of the store proprietor, will uncover *Bittium, Anomia, Odostomia* and other small shells. The bushels of *Venus* can also be examined for specimens of the beautiful form *notata.* Of course exact locality records, except for the obvious "fish store," are harder to come by. But chances are good that, no matter what the owner says, most if not all come from the eastern end of Long Island.

The most interesting shells to be found here are the exotic land snails that are imported in large quantities for the foreign-born epicures of our city. These are displayed in large straw hampers or, in one case, lightly packed in sawdust. Though one can hardly add such species to the faunal lists of the New York City area, it might prove of some interest to the readers of this book to know exactly which ones can be purchased here, especially since these shells are much showier than the rather drab native land snails that actually inhabit our area.

OTALA LACTEA Müller   *Mediterranean Milk Snail*

SIZE: 1 to 1½ inches in diameter, 1 inch high.
DESCRIPTION: Shell whitish, strong, decorated with wide brown bands which may be distinct or indefinite. Sometimes the surface is covered by a myriad of small brownish spots. Lip and aperture a rich chocolate brown. Lip strongly reflected.

This is the commonest snail offered for sale here. It is native to southern Spain and Morocco and has become established in this country in New Orleans, San Antonio and elsewhere in our southern states. Apparently it cannot survive our rather severe winters.

EOBANIA VERMICULATA Müller *Mediterranean Snail*

SIZE: Diameter 1 to 1½ inches, height 1 inch.
DESCRIPTION: Shell similar to *Otala,* but smoother and not quite so strong. The color lines are usually more definite. Lip and aperture always white.

This shell from southern Europe has also become established in isolated colonies in our southern states, where it seems to cause little, if any, harm. Indeed it has become an interesting supplement to local diets. Its white lip and aperture immediately separate it from *Otala*.

HELIX POMATIA Linné *Apple Snail*

SIZE: Diameter 1¾ inches, height 1½ inch.
DESCRIPTION: Shell globose, large, rather thin, with a longitudinally wrinkled surface. Color buff with rather wide, cinnamon brown bands which are sometimes distinct and sometimes ill-defined. Lip thin, unreflected. Aperture very large, edge of the lip covering the umbilicus.

This is the famous apple snail of the ancient Romans and the escargot of the French. It is sometimes sold packaged, with the meat in flat cans, the empty shells in high, transparent plastic containers. A colony of this central European snail is said to be flourishing near Ann Arbor, Michigan. The globose shape and the thin lip identify it immediately.

Sometimes a shell very similar to *pomatia* is seen in the markets. This is known as HELIX MELANOSTOMA Draparnaud and differs in that the lip and the outer portion of the aperture are colored a rich brown (whence its name, which means "black mouth").

HELIX ASPERSA Müller *Speckled Garden Snail*

SIZE: 1½ inch diameter, 1½ inch high.
DESCRIPTION: Shell obliquely globose, thin, with the spire moderately raised; surface wrinkled or otherwise roughened, generally without umbilicus; lip reflected, white. Color yellow, decorated by bands of chestnut brown to

chocolate. These bands are interrupted by yellow flecks and streaks.

This is the harmful speckled garden snail of subtropical and tropical countries. It is now distributed everywhere in such regions, where it frequently becomes the most common snail of all, especially in urban areas. It is a particular pest in the gardens of Los Angeles, where it is known simply as "that goddam snail."

A snail similar in every respect but a good deal larger (about 2 inches high) is known as HELIX ASPERSA MAXIMA Taylor and is also frequently encountered in the markets.

THEBA PISANA Müller *The Small Gray Snail*

SIZE: Height ½ inch, diameter ¾ inch.
DESCRIPTION: Shell somewhat depressed, umbilicate, lip unreflected, simple but thickened slightly within in mature specimens. Color ivory yellow, commonly decorated by brown lines which can be entire or interrupted; sometimes the base is stained by brownish areas.

This very common southern European snail has settled in California where it has become a plant nuisance. It is very pro-

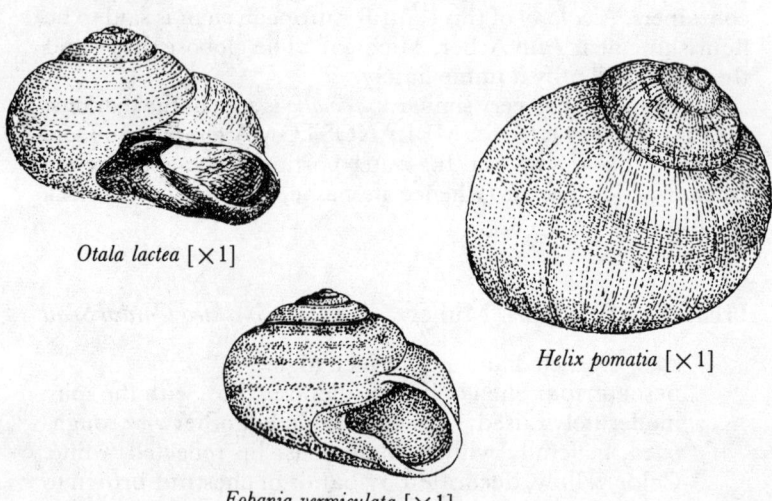

Otala lactea [×1]

Helix pomatia [×1]

Eobania vermiculata [×1]

lific and much money is being spent in combating it. It is a popular European article of diet and because of its small size is not so expensive as *Helix* or *Otala*. However, it is less commonly available in the markets. The immature shell is flattened and provided with a strong peripheral keel.

HELIX APERTA Born *Open Edible Snail*

SIZE: Height 1½ inch, diameter 1 inch.

DESCRIPTION: The shell is very thin, pellucid, shiny, longitudinally wrinkled. The color is yellowish brown, darker at the base of the numerous, rather deep wrinkles. The interior is covered by a much interrupted layer of white. Aperture huge, spire scarcely raised; lip unreflected, very slightly thickened within.

This is the most expensive snail in our markets. It usually arrives in a dormant state, packed in boxes of sawdust. Packed thus, the snail closes its vast aperture by a stout, strongly convex, white epiphragm which it rapidly discards when it is exposed to moisture. Its flavor is somewhat more delicate than that of the other snails.

The snails described above are the most common that are offered for sale. However, other species appear occasionally, for

Helix aspersa [×1]

Theba pisana [×1]

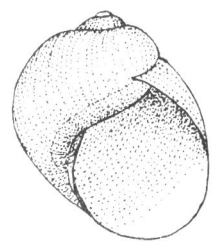

Helix aperta [×1]

any land snail can be made to serve as food. The only problem is to collect them in large enough quantities. As soon as this can be done, the snails are rushed to market and some may even arrive here. Hence it is useful, if one is interested, to make periodic visits to those little centers of exoticism in New York, the Italian fish stores.

A word of precaution may be offered here for those readers who might be inspired to dine on the snails as well as to collect the shells. The snails bought in the markets can, of course, be prepared immediately for the table in any of the many fashions described so succulently in cook books. However, if one ventures to make a meal of snails collected in gardens or forests, it is wiser to wait at least twenty-four hours before eating them, because snails can flourish on plants that are highly poisonous to humans and must be given time to discard any noxious matter that they may have ingested.

Shell Lists
of Special Areas

In compiling the following lists, we have tried to confine ourselves to a few representative localities in the New York City area (see locality map). It must be understood that these lists are not meant to be complete; we mention only the shells we have actually collected there in the course of repeated visits. Similarly no one should hope to find all these species on any single day. The season, the currents, the prevailing winds all play a rôle in casting shells on the beach or depositing them in shallow water. And since such weather elements vary from day to day, a series of regular visits is required before a complete collection of any single locality can be made. It is hoped that these lists will be of some aid to the collector in identifying—or determining, as zoologists say—the species that are found. (The figure following the name of each species refers to the page of this book on which the shell is described.)

CONEY ISLAND

The commonest shells here are *Crepidula fornicata* and *Mytilus edulis*. Coney Island is actually a bay area, and hence outer beach shells like *Spisula solidissima* are by no means as common as they are at Rockaway Beach.

    Anomia simplex, 79
    Brachidontes demissus, 82
    Crassostrea virginica, 76
    Donax fossor, 93
    Ensis directus, 94
    Mulinia lateralis, 96
    Mya arenaria, 99
    Mytilus edulis, 80
    Pandora gouldiana, 83
    Petricola pholadiformis, 100
    Spisula solidissima, 96
    Tellina agilis, 92

    Busycon canaliculatum, 62
    Crepidula convexa, 50
    Crepidula fornicata, 49
    Crepidula plana, 50
    Littorina littorea, 52
    Nassarius obsoletus, 58
    Nassarius trivittatus, 59
    Polinices duplicatus, 48
    Polinices heros, 46

ROCKAWAY BEACH

This list applies also to other open beaches in the western part of Long Island, such as Jones Beach, Point Lookout, Fire Island, Long Beach, etc.

    Anadara pexata, 74
    Anadara transversa, 75
    Anomia simplex, 79 (dead only)
    Arctica islandica, 85 (rarissime!)
    Astarte castanea, 86 (single valves only)
    Astarte undata, 86 (ditto)
    Barnea truncata, 101 (at Far Rockaway and in water-logged wood)
    Corbula contracta, 98
    Cyrtopleura costata, 100 (rare, fragments only)
    Divaricella quadrisulcata, 87 (not uncommon, single valves only)

Donax fossor, 93 (abundant some years, absent others)
Ensis directus, 94 (single valves only)
Gemma gemma, 90 (in shell drifts)
Hiatella arctica, 98 (rare)
Laevicardium mortoni, 88 (rare, single discolored valves only)
Mulinia lateralis, 96 (ditto)
Mytilus edulis, 80 (on jetties)
Noetia ponderosa, 76 (fossil)
Nucula proxima, 73 (rare)
Pecten irradians, 77 (single valves only)
Pecten magellanicus, 78
Petricola pholadiformis, 100
Pitar morrhuana, 89 (live specimens occasionally found)
Siliqua costata, 94 (rare, fragments only)
Spisula solidissima, 96 (abundant, all sizes)
Tagelus plebeius, 94 (single valves only)
Tellina agilis, 92 (occasionally abundant)
Teredo navalis, 102 (in water-logged wood)
Venericardia borealis, 87 (single valves only, rare)
Venus mercenaria, 88 (mainly single valves)
Zirfaea crispata, 102 (in water-logged wood)

Anachis avara, 56 (rare)
Buccinum undatum, 60 (dead only, very rare)
Busycon canaliculatum, 62
Busycon carica, 63
Cerithiopsis greenii, 66 (dead only, in shell drift)
Colus islandicus, 61 (dead only, very rare)
Crepidula convexa, 50
Crepidula fornicata, 49
Crepidula plana, 50 (common inside Busycon shells)
Crucibulum striatum, 51 (very rare)
Cylichna oryza, 70 (in shell drift)
Epitonium humphreysii, 46 (rare)
Epitonium multistriatum, 46 (rare)
Eupleura caudata, 55
Lacuna vincta, 54 (on sea weed)
Mangelia *species*, 69 (in shell drift)
Mitrella lunata, 57 (on sea weed)
Nassarius obsoletus, 58 (not common on beach)
Nassarius trivittatus, 59 (usually dead)
Nassarius vibex, 60 (rare)
Polinices duplicatus, 48
Polinices heros, 46
Polinices triseriatus, 48 (rare, dead only)

Retusa canaliculata, 69 (in shell drift)
Seila adamsi, 66 (ditto)
Triphora perversa nigrocincta, 67 (ditto)
Turbonilla *species*, 68 (ditto)
Urosalpinx cinerea, 54

HAMILTON BEACH

This is a beach located on the northern shore of Jamaica Bay and as such is typical of other Jamaica Bay areas, such as Roxbury, Broad Channel, Riis Park, Canarsie, Bergen Beach, etc.

Brachidontes demissus, 82 (very common in intertidal beds)
Crassostrea virginica, 76
Ensis directus, 94 (common)
Gemma gemma manhattensis, 90 (very common)
Laevicardium mortoni, 88 (not rare)
Macoma balthica, 92
Mulinia lateralis, 96
Mya arenaria, 99 (abundant)
Mytilus edulis, 80
Nucula proxima, 73 (rare)
Pecten irradians, 77 (single valves)
Petricola pholadiformis, 100
Solemya velum, 73
Tagelus plebeius, 94
Tellina agilis, 92 (white variety only)
Venus mercenaria, 88 (common)

Busycon canaliculatum, 62
Crepidula convexa, 50 (common)
Crepidula fornicata, 49 (common)
Eupleura caudata, 55
Haminoea solitaria, 64 (seasonal)
Hydrobia minuta, 68 (on sea lettuce)
Littorina littorea, 52 (very common intertidally)
Littorina obtusata, 53 (on *Fucus* sea weed)
Littorina saxatilis, 53 (on pilings)
Melampus bidentatus, 64 (cόmmon intertidally)
Nassarius obsoletus, 58 (the most abundant snail in the area)
Odostomia trifida, 67 (common on dead shells, stones, etc.)
Ovatella myosotis, 65 (on pilings above highwater)
Polinices duplicatus, 48 (ditto)
Polinices heros, 46 (occasionally, dead)
Urosalpinx cinerea, 54 (common)

SHELL LISTS OF SPECIAL AREAS · 115

WARD POINT, TOTTENVILLE, STATEN ISLAND

This is an unusually rich collecting spot for the New York City area. Many species that are rare elsewhere are common here. It is basically a bay area with a fauna that has miraculously not been decimated by pollution.

Barnea truncata, 101
Brachidontes demissus, 82
Crassostrea virginica, 76
Cyrtopleura costata, 100 (fragments in dredgings)
Ensis directus, 94
Gemma gemma manhattensis, 90
Laevicardium mortoni, 88
Macoma balthica, 92
Mulinia lateralis, 96 (here abundant)
Mya arenaria, 99 (abundant)
Mytilus edulis, 80
Petricola pholadiformis, 100
Tellina agilis, 92 (usually white)
Venus mercenaria, 88 (*plus* notata)

Bittium alternatum, 66 (in dredgings)
Crepidula convexa, 50 (common)
Crepidula fornicata, 49 (ditto)
Crepidula plana, 50 (rare here)
Epitonium humphreysii, 46 (rare)
Epitonium rupicola, 45 (found only here in fair numbers)
Eupleura caudata, 55
Haminoea solitaria, 64
Hydrobia minuta, 68
Littorina littorea, 52
Littorina saxatilis, 53
Mangelia plicosa, 69 (in dredgings)
Nassarius obsoletus, 58 (abundant)
Nassarius trivittatus, 59
Nassarius vibex, 60 (rare, usually dead)
Odostomia trifida, 67
Polinices duplicatus, 48 (ditto)
Polinices heros, 46 (dead)
Retusa canaliculata, 69 (common in dredgings)
Urosalpinx cinerea, 54 (very common, very varied in color)

BAYVILLE, LONG ISLAND

This area is typical of the sandy beaches on the North Shore of Long Island, such as Wading River, Sunken Meadow, Huntington, Orient Point, etc. Cold Spring Harbor is a classic collecting locality in this region, but few species can be collected without dredging. Bayville is ideally located because it has the Sound on one side, a bay on the other, large stretches of sandy beach, and a series of rocks towards the east. This provides many habitats for various types of mollusks.

    Anadara pexata, 74
    Anadara transversa, 75
    Anomia simplex, 79 (very common)
    Brachidontes demissus, 82 (common in bay)
    Crassostrea virginica, 76
    Ensis directus, 94
    Gemma gemma, 90 (common in Sound)
    Gemma gemma manhattensis, 90 (common in bay)
    Laevicardium mortoni, 88
    Mulinia lateralis, 96
    Mya arenaria, 99
    Mytilus edulis, 80
    Nucula proxima, 73
    Pecten irradians, 77
    Petricola pholadiformis, 100
    Solemya velum, 73 (in bay)
    Spisula solidissima, 96 (not common here)
    Tellina agilis, 92
    Venus mercenaria, 88

    Acmaea testudinalis, 44 (common on rocks below low water mark)
    Bittium alternatum, 66
    Busycon canaliculatum, 62
    Busycon carica, 63
    Crepidula convexa, 50
    Crepidula fornicata, 49
    Crepidula plana, 50
    Epitonium humphreysii, 46 (rare)
    Littorina obtusata, 53
    Littorina saxatilis, 53
    Mangelia plicosa, 69
    Nassarius obsoletus, 58 (abundant in bay)
    Nassarius trivittatus, 59 (common in sand in shallow water near the rocks)

Nassarius vibex, 60 (occasionally, dead)
Polinices duplicatus, 48
Polinices heros, 46
Polinices triseriatus, 48 (the unbanded form is common in shallow water near the rocks)

SPARKILL, ROCKLAND COUNTY

Although land and fresh water shells can be found in every likely spot in our area, this locality is the single richest spot for these shells that we know of. It is located between Valentine and Ferdun Avenues, and is conspicuously marked by Upper Ferdun Pond of Sparkill Creek. Here it is spanned by a high steel bridge that supports the upper roadway of route 9W. It is located about 13 miles north of the New Jersey terminus of the George Washington Bridge.

Fresh Water Shells

Amnicola limosa, 37
Helisoma anceps, 29
Lymnaea palustris elodes, 26
Musculium *species*, 42
Physa heterostropha, 31
Pisidium variabile, 42
Planorbula jenksii, 29
Promenetus hudsonicus, 30
Sphaerium rhomboideum, 42
Valvata tricarinata, 37

Land Shells

Anguispira alternata, 9
Cionella lubrica, 18
Discus cronkhitei, 15
Gastrocopta contracta, 22
Hawaiia minuscula, 24
Helicodiscus parallelus, 17
Mesodon thyroidus, 5
Oxychilus draparnaldi, 15
Retinella electrina, 14
Retinella indentata, 14
Stenotrema hirsutum, 7
Succinea ovalis, 10
Triodopsis albolabris, 3
Triodopsis tridentata, 4
Vallonia pulchella, 19
Zonitoides arboreus, 12
Zonitoides nitidus, 13

Glossary

Acute	—	sharply pointed
Aestivate	—	to pass the summer in an inactive state
Anterior	—	the forward or head end; opposite of posterior
Aperture	—	the opening of the shell; the entrance to the body whorl of snails
Apex	—	the tip of the spire in snail shells
Aquatic	—	pertaining to or living in water
Axis	—	the center around which the whorls coil
Base	—	the extremity opposite the apex in snail shells
Beak	—	see umbo
Bivalve	—	a clam or pelecypod mollusk; a shell with two valves
Body whorl	—	the last and largest whorl of a snail shell
Byssus	—	the threads by which certain bivalves attach themselves to solid objects

Canal	—	a grooved projection of the lip of the aperture in many snail shells
Carina	—	a keel-like ridge (carinae, plural)
Columella	—	the axial pillar around which the whorls of the snail shell coil, visible at inner lip of aperture
Commensalism	—	the association of two or more individuals of different species in which one or more is benefited and the others are not harmed
Conchology	—	the study of molluscan shells; see also Malacology
Conic	—	shaped like a cone
Crenulate	—	finely notched or delicately corrugated as on the edge of some bivalve shells, or the outer lip of some snail shells
Dextral	—	right-handed; having the aperture on the right side of a snail shell when the apex is held upward
Dorsal	—	toward or pertaining to the upper surface or the back
Epiphragm	—	a thin sheet of dried mucus secreted by land shells closing their aperture to prevent loss of heat and moisture during aestivation or hibernation
Excavated	—	hollowed out, as the area around umbilicus in some snail shells
Fauna	—	the animal life living in a given region or during a specified period of time
Fusiform	—	spindle-shaped, *i.e.* swelling in the central part, tapering at extremities
Gastropods	—	univalve shells, such as snails, slugs, nudibranchs, etc.
Hibernate	—	to pass the winter in an inactive state
Hinge	—	interlocking tooth devices of the valves of bivalve shells
Hyaline	—	glossy, or semitransparent
Incised	—	sculptured with depressed lines or grooves
Inner lip	—	the portion of aperture near the basal part of the columella

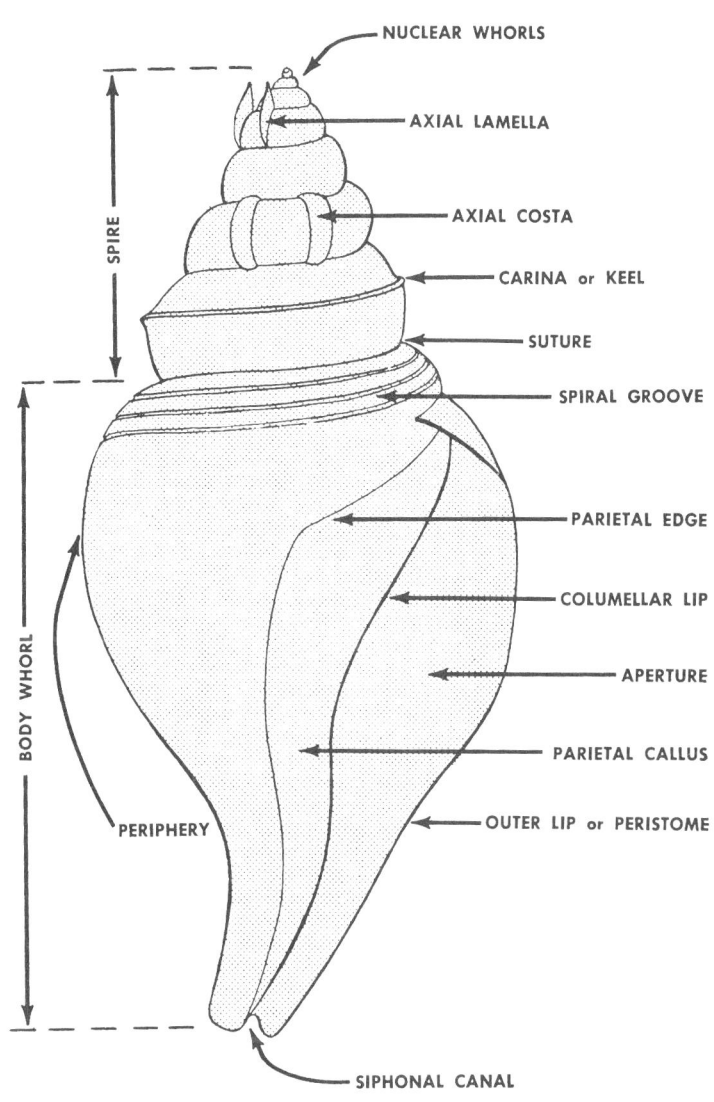
Composite figure illustrating the parts of a GASTROPOD (UNIVALVE)

Lamella	—	narrow longitudinal ridge; usually inside the aperture (lamellae, plural)
Lamellibranchia	—	another name for the bivalve shells (pelecypods)
Ligament	—	cartilage and elastic resilium connecting the valves of pelecypods at the umbo
Lips	—	the margins of the aperture of snail shells
Malacology	—	the study of mollusks concerning the soft anatomy as well as the shell
Malleated	—	appearing as though hammered
Mantle	—	membrane which encloses the molluscan viscera and secretes the shell of mollusks
Matte	—	lusterless, dull surface texture
Mollusks	—	animals belonging to the phylum Mollusca, including gastropods, pelecypods, cephalopods, scaphopods, and chitons (also spelled molluscs)
Operculum	—	a horny or shelly plate which closes the aperture in some snails when the animal retracts within its shell
Ocelli	—	tiny, simple eyes, as in the scallop
Outer lip	—	outer edge of the aperture of snail shells
Ovate	—	egg-shaped
Pallet	—	a structure in shipworm pelecypods used to block the entrance to their burrows
Pallial line	—	the impression or scar on the inner surface of the bivalve shell, marking the attachment of the mantle
Pallial sinus	—	an indentation in the pallial line
Parietal callus	—	enamel on the inside wall of shell just within the aperture of snails
Parietal wall	—	inside wall of shell within the aperture of snails, nearest the columella
Peduncle	—	a stalk or stem-like structure
Pelecypods	—	bivalved mollusks, including clams, oysters, etc.
Pellucid	—	transparent or clear
Penultimate whorl	—	the whorl before the last or body whorl
Periostracum	—	the skin-like outer covering of many shells
Periphery	—	the edge or boundary of an area
Peristome	—	the outer lip of the aperture of a snail shell
Posterior	—	the hinder part, or towards the rear end; opposite of anterior
Process	—	a projection

Composite figure illustrating the parts of a PELECYPOD (BIVALVE)

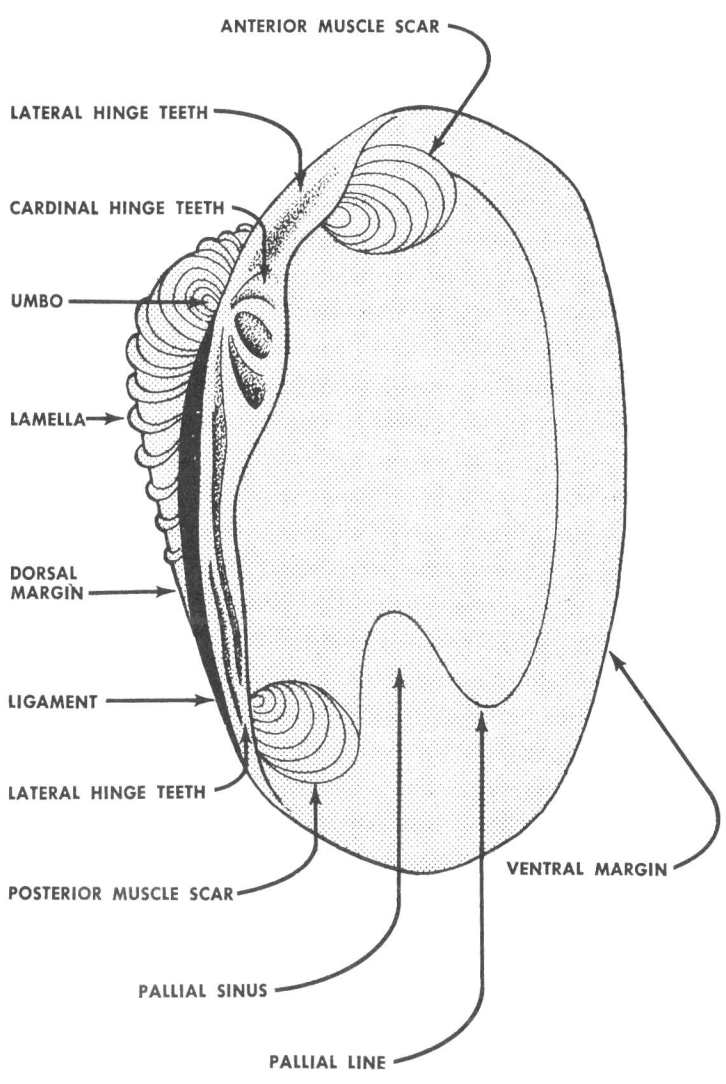

Quadrate	—	rectangular in shape
Radula	—	the dental apparatus possessed by most snails and all other mollusks except pelecypods, composed of a ribbon to which are fixed numerous plates or teeth
Reflected	—	bent over and backward, as in the outer lip of some snail shells
Resilium	—	the internal cartilage in the ligament of some clams
Serrate	—	having notches or projections suggesting teeth of a saw
Sinistral	—	left-handed; having the aperture on the left side of a snail shell when the apex is held upward
Sinus	—	an indentation or deeply cut cavity
Siphon	—	a tube-like extension of the mantle by which water enters or leaves the mantle cavity
Spire	—	the upper whorls, the coils above the body whorl of a snail shell
Striae	—	very fine lines (stria, singular)
Suture	—	the junction between whorls of a snail shell
Teeth	—	dentate-like nodules resembling teeth, as in the aperture of some snails, or in the hinge of most bivalves; also the chitinous plates of the gastropod radula
Terrestrial	—	living on land
Trigonal	—	triangular or three-cornered in shape
Umbilicus	—	a small hollow usually at the center of the base of the body whorl in some snail shells
Umbo	—	the earliest part, or beak of the bivalve shell
Varices	—	prominently raised vertical ridges on the surface of snail shells originally formed at aperture (varix, singular)
Ventral	—	toward the lower side or belly; away from the back
Whorl	—	one complete spiral turn (volution) of the spire, as in most snail shells

Taxonomy of the Mollusks

Although many scholars have worked on the problem, an absolute system of arrangement of the mollusks in a natural order is still a matter of much disagreement. Especially in the matter of the bivalves is there an astounding amount of controversy and indecision. In the following tables, the authors followed, as far as possible, the systems established by reputable workers who had taken the trouble to work out their ideas completely and did not merely content themselves with suggested modifications in the work of others. For these reasons Dr. Johannes Thiele is our guide, with few modifications, for the marine and fresh water shells, and Dr. Henry A. Pilsbry for the land mollusks. The solid labor of these scholars and their intricately worked out systems will undoubtedly remain the most generally accepted standard for years to come.

CLASS and SUBCLASS	ORDER and SUBORDER	SUPERFAMILY
AMPHINEURA		
GASTROPODA		
Prosobranchia	Archaeogastropoda	Patellacea
	Mesogastropoda	Architaenioglossa
		Valvatacea
		Littorinacea
		Rissoacea
		Cerithiacea
		Ptenoglossa
		Calyptraeacea
		Naticacea
	Neogastropoda	Muricacea

TAXONOMY OF THE MOLLUSKS · 127

FAMILY and SUBFAMILY			PAGE NUMBER
Ischnochitonidae	M*	Chaetopleura apiculata	103
Acmaeidae	M	Acmaea testudinalis	44
Viviparidae	F	Viviparus malleatus	34
	F	Viviparus contectoides	35
	F	Campeloma decisum	36
Valvatidae	F	Valvata tricarinata	37
Lacunidae	M	Lacuna vincta	54
Littorinidae	M	Littorina littorea	52
	M	Littorina obtusata	53
	M	Littorina saxatilis	53
	M	Littorina irrorata	52
Hydrobiidae	M	Hydrobia minuta	68
	F	Amnicola limosa	37
	F	Bulimus tentaculatus	36
Melaniidae	F	Goniobasis virginica	34
Cerithiidae	M	Bittium alternatum	66
Cerithiopsidae	M	Cerithiopsis greenii	66
	M	Seila adamsi	66
Triphoridae	M	Triphora perversa nigrocincta	67
Epitonidae	M	Epitonium multistriatum	46
	M	Epitonium angulatum	46
	M	Epitonium rupicola	45
	M	Epitonium humphreysii	46
Calyptraeidae	M	Crucibulum striatum	51
	M	Crepidula plana	50
	M	Crepidula convexa	50
	M	Crepidula fornicata	49
Naticidae	M	Polinices heros	46
	M	Polinices duplicatus	48
	M	Polinices triseriatus	48
Muricidae	M	Urosalpinx cinerea	54
	M	Eupleura caudata	55
	M	Thais lapillus	56

*M = Marine; F = Fresh water; L = Land

128 · SHELLS OF THE NEW YORK CITY AREA

CLASS and SUBCLASS	ORDER and SUBORDER	SUPERFAMILY
Prosobranchia (*continued*)	Neogastropoda (*continued*)	Buccinacea
		Toxoglossa
Opisthobranchia	Pleurocoela	Cephalaspidea
		Pyramidellacea
Pulmonata	Basommatophora	Actophila
		Hygrophila

FAMILY and SUBFAMILY			PAGE NUMBER
Columbellidae	M	Anachis avara	56
	M	Mitrella lunata	57
Buccinidae	M	Buccinum undatum	60
	M	Colus islandicus	61
	M	Colus stimpsoni	62
	M	Colus stonei	62
	M	Busycon canaliculatum	62
	M	Busycon carica	62
Nassariidae	M	Nassarius obsoletus	58
	M	Nassarius trivittatus	59
	M	Nassarius vibex	60
Turridae	M	Mangelia *species*	69
Acteonidae	M	Acteon punctostriatus	70
Atyidae	M	Haminoea solitaria	64
Retusidae	M	Retusa canaliculata	69
Scaphandridae	M	Cylichna oryza	70
Pyramidellidae	M	Turbonilla *species*	68
	M	Odostomia *species*	67
Ellobiidae	L	Carychium exiguum	20
	L	Carychium exile	21
	M	Ovatella myosotis	65
	M	Melampus bidentatus	64
Physidae	F	Aplexa hypnorum	32
	F	Physa heterostropha	31
	F	Physa ancillaria	32
Lymnaeidae	F	Lymnaea palustris elodes	26
	F	Lymnaea columella	26
	F	Lymnaea auricularia	26
	F	Lymnaea humilis	27
Planorbidae	F	Helisoma trivolvis	28
	F	Helisoma anceps	29
	F	Planorbula jenksii	29
	F	Gyraulus parvus	30
	F	Promenetus hudsonicus	30
Ancylidae	F	Ferrissia fusca	32
	F	Ferrissia parallela	33

CLASS and SUBCLASS	ORDER and SUBORDER	SUPERFAMILY
Pulmonata (*continued*)	Stylommatophora	
	Sigmurethra	
	Orthurethra	

FAMILY and SUBFAMILY			PAGE NUMBER
Helicidae	L	Cepaea nemoralis	11
Polygyridae			
Polygyrinae	L	Stenotrema hirsutum	7
	L	Stenotrema fraternum	7
	L	Mesodon thyroidus	5
Triodopsinae	L	Triodopsis tridentata	4
	L	Triodopsis albolabris	3
	L	Triodopsis denotata	4
Haplotrematidae	L	Haplotrema concavum	8
Zonitidae			
Euconulinae	L	Euconulus fulvus	22
	L	Euconulus chersinus	22
Zonitinae	L	Retinella electrina	14
	L	Retinella indentata	14
	L	Oxychilus cellarius	14
	L	Oxychilus draparnaldi	15
	L	Oxychilus alliarius	15
	L	Hawaiia minuscula	24
Gastrodontinae	L	Ventridens ligera	6
	L	Zonitoides nitidus	13
	L	Zonitoides arboreus	12
Endodontidae			
Endodontinae	L	Anguispira alternata	9
	L	Discus cronkhitei	15
	L	Discus rotundatus	16
Helicodiscinae	L	Helicodiscus parallelus	17
Punctinae	L	Punctum minutissimum	24
Strobilopsidae	L	Strobilops labyrinthica	23
	L	Strobilops aenea	23
	L	Strobilops affinis	23
Pupillidae			
Gastrocoptinae	L	Gastrocopta *species*	22
Vertigininae	L	Vertigo *species*	22
	L	Columella edentula	22

CLASS and SUBCLASS	ORDER and SUBORDER	SUPERFAMILY
Pulmonata *(continued)*	Stylommatophora Orthurethra *(continued)*	
	Heterurethra	
BIVALVIA (Pelecypoda)	Taxodonta	Nuculacea
		Arcacea
	Anisomyaria	Mytilacea
		Pectinacea
		Anomiacea
	Eulamellibranchiata Schizodonta	Unionacea
	Heterodonta	Astartacea
		Carditacea
		Sphaeriacea
		Cyprinacea

FAMILY and SUBFAMILY			PAGE NUMBER
Valloniidae	L	Vallonia pulchella	19
	L	Vallonia costata	20
	L	Vallonia excentrica	20
Cionellidae	L	Cionella lubrica	18
Succineidae	L	Succinea ovalis	10
	L	Succinea avara	10
Nuculidae	M	Nucula proxima	73
Nuculanidae	M	Yoldia limatula	74
	M	Yoldia sapotilla	74
Solemyidae	M	Solemya velum	73
Arcidae	M	Anadara pexata	74
	M	Anadara transversa	75
	M	Noetia ponderosa	76
Mytilidae	M	Mytilus edulis	80
	M	Modiolus modiolus	82
	M	Brachidontes demissus	82
Pectinidae	M	Pecten irradians	77
	M	Pecten magellanicus	78
Anomiidae	M	Anomia simplex	79
	M	Anomia aculeata	80
Ostreidae	M	Crassostrea virginica	76
Margaritanidae	F	Margaritana margaritifera	40
Unionidae	F	Anodonta cataracta	38
	F	Anodonta implicata	38
	F	Elliptio complanatus	39
Astartidae	M	Astarte undata	86
	M	Astarte castanea	86
Carditidae	M	Venericardia borealis	87
Sphaeriidae	F	Sphaerium *species*	42
	F	Musculium *species*	42
	F	Pisidium *species*	42
Arcticidae	M	Arctica islandica	85

CLASS and SUBCLASS	ORDER and SUBORDER	SUPERFAMILY
	Eulamellibranchiata Heterodonta (*continued*)	Dreissenacea
		Lucinacea
		Cardiacea
		Veneracea
		Mactracea
		Tellinacea
	Adapedonta	Solenacea
		Saxicavacea
		Myacea
		Adesmacea
	Anomalodesmata	Pandoracea
CEPHALOPODA		
Dibranchia	Decapoda	Sepiacea

FAMILY and SUBFAMILY			PAGE NUMBER
Dreissenidae	M	Congeria leucophaeata	83
Lucinidae	M	Divaricella quadrisulcata	87
Cardiidae	M	Laevicardium mortoni	88
Veneridae	M	Venus mercenaria	88
	M	Gemma gemma	90
	M	Gemma purpurea	90
	M	Pitar morrhuana	89
Petricolidae	M	Petricola pholadiformis	100
Mesodesmatidae	M	Mesodesma arctatum	97
Mactridae	M	Mulinia lateralis	96
	M	Spisula solidissima	95
Donacidae	M	Donax fossor	93
Tellinidae	M	Tellina agilis	92
	M	Macoma balthica	92
	M	Macoma calcarea	92
Solenidae	M	Ensis directus	94
	M	Tagelus plebeius	94
	M	Tagelus divisus	94
	M	Siliqua costata	94
Hiatellidae	M	Hiatella arctica	98
Corbulidae	M	Corbula contracta	98
Myidae	M	Mya arenaria	99
Pholadidae	M	Cyrtopleura costata	100
	M	Barnea truncata	101
	M	Zirfaea crispata	102
Teredinidae	M	Teredo navalis	102
Lyonsiidae	M	Lyonsia hyalina	84
Pandoridae	M	Pandora gouldiana	83
Periplomatidae	M	Periploma leanum	84
Spirulidae	M	Spirula spirula	70

Supplement of Additional Species

THE LAND SNAILS

Depressed Belly-tooth Snail
VENTRIDENS SUPPRESSUS Say

SIZE: about 1/4 inch in diameter.
DESCRIPTION: The shell is much depressed, the spire but little raised. It is pale yellowish amber in color, the body whorl with an opaque whitish splotch near the aperture. It has a small umbilicus and in the adult stage a narrow but rather high tooth inside the base of the aperture.

This relative of *V. ligera* is smaller and more depressed. It is most easily recognized by the narrow but prominent tooth just inside the aperture. In its immature stages the shell looks quite different, since there are three to five small denticles along the interior margin of the outer lip. In maturity, however, only the single basal tooth remains. The species is not rare around Peekskill and it has also been reported from Staten Island—presumably before the Verrazano Bridge was built.

MESOMPHIX INORNATUS Say   *Plain Ground-Snail*

SIZE: up to ¾ inch in diameter, about ¼ inch high.
DESCRIPTION: The shell is low, somewhat olivaceous brown, moderately glossy, rather thin and fragile. There is a small umbilicus and the lip is thin and unexpanded.

This species is about the same size as its relative *Oxychilus draparnaldi*, but it is decidedly higher, less glossy and different in color, and has a much narrower umbilicus. A native of North America, it is found in woods far from any town, whereas *O. draparnaldi* is a post-Columbian immigrant and rarely leaves the immediate haunts of men in towns and cities.

Copper Ground-Snail
MESOMPHIX CUPREUS Rafinesque

This species is similar to *M. inornatus* but differs as follows: it is generally larger and higher, the umbilicus is proportionately larger and the color is darker, in some specimens like dull copper. This color probably gave rise to its Latin name. It was found recently in the Wachung Reservation and elsewhere in northern New Jersey. It seems to be more of a forest lover than *M. inornatus*.

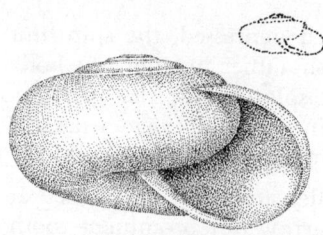

Mesomphix inornatus [×3]

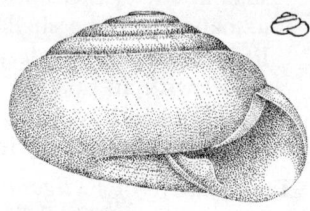

Ventridens suppressus [×6]

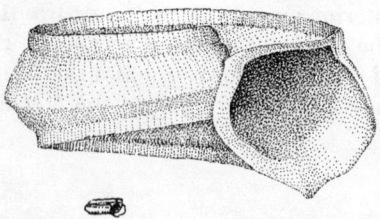

Valvata bicarinata [×12]

MOLLUSKS OF FRESH WATER

Bell-mouthed Rams Horn
HELISOMA CAMPANULATUM Say

This species looks like the Three-whorled Rams Horn (*H. trivolvis*) but differs strikingly in having the last whorl expanded into a large, bell-shaped aperture, hence its Latin name. In addition the whorls are much narrower and more tightly coiled. It is a common shell in upstate New York but recently has been taken near Lake Hopatcong in New Jersey.

Dilated Rams Horn
MICROMENETUS DILATATUS Gould

This species of small planorbid is about as large as the Tiny Rams Horn (*Gyraulus parvus*). However, it differs in having a keel at the very top of the shell and in having the last whorl proportionately wider. This probably suggested its Latin name, which means "dilated." The color, when the animal is taken alive, is light reddish brown.

This species was recently taken at the table from some watercress which formed part of the salad. It is also found, but not commonly, in the streams of the eastern part of Long Island in Southold, Greenport and Noyack.

VALVATA BICARINATA Say   *Two-keeled Valve Snail*

Very similar in size, color, and shape to *V. tricarinata*. It differs, as the name indicates, in having only two keels or carinae. This species was recently found in a small pond near Bayside, in Queens County. It is the only *Valvata* species ever reported from Long Island and is undoubtedly a recent introduction. It shares its home in Queens with the other stranger, *Viviparus georgianus* (formerly called *Viviparus contectoides*), which also only recently began to show up within the city limits.

MARINE GASTROPODS

POLINICES IMMACULATUS Totten   *Spotless Moon*

This is a very small moon snail about 3/8 inch high, milk white and moderately glossy. The umbilicus is open. It lives in deeper water but is sometimes washed ashore or found in fish stomachs.

NATICA PUSILLA Say   *Miniature Moon*

This tiny moon snail differs from the other moon snails in our area by having a calcareous operculum instead of a thin horny one. It resembles the Spotless Moon (*Polinices immaculatus*) in size but is normally ornamented with faint brown spots. The umbilicus, moreover, is completely closed.

Little Awl Horn
CERITHIOPSIS SUBULATA Montagu

This shell resembles *C. greenii* in shape and sculpture, but it is much larger, reaching almost 3/4 inch. The second part of the name means "awl-like" in Latin, since *C. subulata* much resembles that tool. It is said to live in 1 to 43 fathoms of water off shore, but records from the New York City area are very rare.

ASSIMINEA MODESTA H. C. Lea   *Simple Assiminea*

SIZE: about 3/8 inch in height.
DESCRIPTION: The shell is light chestnut in color, translucent but solid. When the animal is collected alive, a line appears below the suture, making the suture look doubled, as indicated by the illustration. The umbilicus is very narrow and the operculum is thin, dark, and has only a few wide whorls.

This snail is easily recognized in spite of its small size by the light brown color and the moderately glossy sheen of its shell, as well as the "doubled" suture. We found it in fair numbers in cracks in the jetties in Jamaica Bay near Far Rockaway, where it lived together with *Ovatella myosotis*. When the shell is taken alive, its bright red buccal mass makes it even easier to identify. In addition, the tentacles each bear two eye spots, not one.

Francis Noyes Balch, a nineteenth-century malacologist, wrote of the scarcity of this species near Newport, Rhode Island, near the northern limit of its range. He reports: "After collecting all that could be found at one time, on the next day about the same number would appear on the same space of five feet by three, and nowhere else." The shells we found in Far Rockaway were not so exclusive.

Handsome Gut Snail
CAECUM PULCHELLUM Stimpson

This species has a unique shell. It looks like a tiny, 1/16-inch long, curved tube which is ornamented with a series of closely set axial rings. It is generally chalky white in color and is thus hard to detect in the sand in which it lives. When young it has a minute coiled shell which it discards as it matures. It lives in sand and shallow water and has been reported from Gardiners Bay.

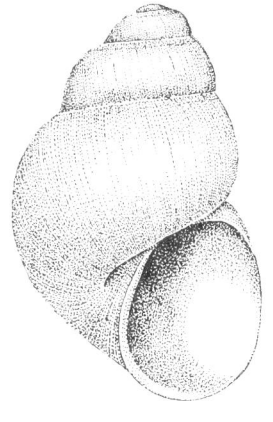

Assiminea modesta [×10]

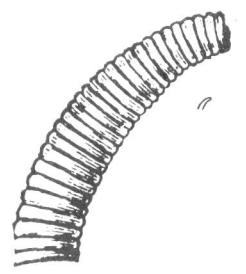

Caecum pulchellum [×15]

MARINE PELECYPODS

MYSELLA PLANULATA Stimpson — *Flat Lepton*

This tiny bivalve is generally about ⅛ inch long. It is more fragile than the Gem Clam (*Gemma gemma*) and less glossy. The beak is to one side, making the shell look like a tiny Wedge Clam (*Donax*). It is white or reddish brown in color and lives attached to eel grass and wharves.

MACOMA TENTA Say — *Narrow Macoma*

SIZE: ½ to ¾ inch in length.
DESCRIPTION: The shell has fragile, elongate white valves. Occasionally there is a faint opalescent tinge in the interior. The narrower, posterior end is slightly twisted to the left.

This common species can easily be mistaken for a young *Mya,* but the absence of the internal shelf near the umbo and the presence instead of two small teeth immediately serve to distinguish it. It is lower and more elongate than either *M. balthica* or *M. calcarea*. It is a common shell on muddy bay bottoms in shallow water.

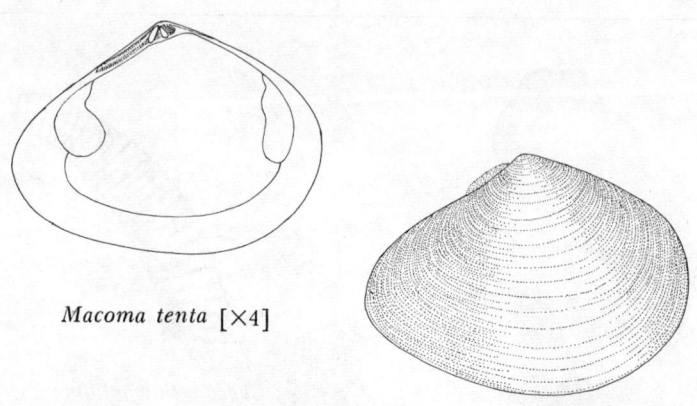

Macoma tenta [×4]

Macoma tenta [×4]

Incised Wedge Clam
CUMINGIA TELLINOIDES Conrad

As the second part of the name of this species indicates, this shell looks like a *Tellina,* or Wedge Clam, especially the Baltic Macoma. It is about ½ inch long and fairly thin, and has the generally oblong shape of the Wedge Clams. It is most easily recognized by its chalky white color and especially by the strong, sharp, closely set concentric lines which cover the surface. It is said to be fairly common, but in Long Island Sound generally only dead shells are found.

SOLEN VIRIDIS Say — *Green Razor*

This clam is much smaller than the Razor Clam (*Ensis directus*) and has a straight rather than curved outline. The color under the periostracum is faintly greenish. It is excessively rare in the northern part of its range.

Bibliography

There are a number of excellent works that deal with the marine mollusks of the Eastern Coast of the United States, including those of the New York area. *Field Guide to the Shells of Our Atlantic Coast* by Percy A. Morris (Houghton, revised 1951) is well illustrated, handy and small, and very reasonable in price. Many of our larger marine shells are discussed in it. *East Coast Marine Shells* by Maxwell Smith (Ann Arbor: 1937 and several revised editions) is a larger work dealing with the same fauna and including many of the minute shells. *Johnsonia,* published by the Museum of Comparative Zoology of Harvard University, is a series of monographs dealing with single families or genera of marine shells. These papers are most thorough discussions of the mollusks involved and all are superbly illustrated. They are absolutely the last word in their field. About 100 genera have been treated to date and a few of these include discussion of mollusks from the New York area. The price is moderately reasonable. *List of Marine Mollusca of the Atlantic Coast from Labrador to Texas* by Charles W. Johnson (Proceedings of the Boston Society of Natural History, Vol. 40 number 1: 1934) is the most complete check list of our marine shells thus far undertaken. It includes the range of each species and the depth at which it is found, and hence is particularly valuable for running identifications down, especially if one knows the locality and habitat of an unnamed shell. This list is out of print but is available at libraries.

In 1954 Dr. R. Tucker Abbott of the Pilsbry Chair of Malacology in the Academy of Natural Sciences of Philadelphia published *American Seashells* (Van Nostrand and Co., New York). This work deals with the marine shells from all coasts of the United States and hence gives a good portion of its pages to the description of our Atlantic shells. This is a magnificently illustrated work, completely authoritative and in many ways a leader in the field. The introduction deals thoroughly with many facets of molluscan biology as well as history, suggestions for collecting, etc.

Our terrestrial shells are included in the monumental *Land Mollusca of North America* by Dr. Henry A. Pilsbry (Academy of Natural Sciences of Philadelphia, 1939-1948) in two huge volumes consisting of two parts each. This is one of the most thorough and valuable works on mollusks ever written. *Field Book of Illinois Land Shells* by Dr. Frank Collins Baker (Urbana, Illinois, 1939) is a small, moderately priced, well-illustrated handbook that includes discussions of most of our land shells.

In contrast to the land and marine shells of the New York area, our fresh-water shells have not been thoroughly treated in recent works on conchology. They are briefly discussed and illustrated in *The Mollusks of the Niagara Frontier Region* by Imogene C. S. Robertson and Clifford L. Blakeslee (Bulletin of the Buffalo Society of Natural Science, Vol. 19 number 3, Buffalo, 1948), a work which also includes short descriptions of most of our land mollusks.

Report on the Invertebrates of Massachusetts by Augustus A. Gould, edited by W. G. Binney (Boston, 1870) is a splendidly illustrated and written work that discusses the land, fresh-water and marine mollusks of the northeastern part of our country. The nomenclature is largely obsolete, but there is little confusion and the work is still very usable. Though long out of print, it is frequently offered for sale at reasonable prices. As can be seen, the authors of the present report leaned heavily on this authoritative work. The only work dealing with New York shells particularly is *Zoology of New York or the New York Fauna Part V. the Mollusca* by James E. De Kay (Albany 1843) which contains many hand colored plates. It is an interesting work that is still occasionally offered for sale. *Check List of the Mollusca of New York* by Elizabeth J. Letson (Bulletin 88 of the New York State Museum, Albany: 1905) is the most extensive list of New York land, fresh-water and marine mollusks. Though the names in many cases are out-dated, the list is particularly valuable for its inclusion of the synonyms by which many of our local shells have been known in older works on conchology.

The Shell Book by Julia E. Rogers (Doubleday Doran, 1951), with the nomenclature revised by Dr. Harald A. Rehder of the United States National Museum, deals with world wide shells, and has several New York shells illustrated and described. There are some valuable chapters on collecting and housing of shell collections. The same is true of *Handbook for Shell Collectors* by Walter Freeman Webb (Rochester: 1936 and several revised editions), which contains much general conchological information. *West Coast Shells* by Josiah Keep, revised by Dr. Joshua L. Bailey, Jr. (Stanford University Press, 1935), though it deals only with species not found in our area, has a splendid chapter on "Scientific Names and What They Tell Us," a simple introduction to the reasons for and problems associated with nomenclature. In addition the book is written with considerable charm that will serve to whet the appetite of the beginner.

The Nautilus is a quarterly journal devoted to the interests of conchologists. It was begun in 1886 and at present is published in Philadelphia by Drs. R. Tucker Abbott, H. Burrington Baker and Charles B. Wurtz. It often contains articles dealing with our local shells and in general is a fascinating storehouse of conchological lore.

The following is a list of papers dealing exclusively with the mollusks of the New York area:

Smith, Sanderson. 1859. "Depth [sic] of Mollusks of Peconic and Gardiner's Bay, Long Island, New York." *American Journal of Science*, vol. 27, p. 281–283.

Smith, Sanderson. 1862. "On the Mollusca of Peconic and Gardiner's Bay, Long Island, New York." *Annals of the Lyceum of Natural History of New York*, vol. 7, p. 147–168.

Hubbard, J. W. and Sanderson Smith. 1865. "Catalogue of Mollusca of Staten Island, N. Y." *Annals of the Lyceum of Natural History of New York,* vol. 8, p. 151–154.

Smith, Sanderson. 1865. "Catalogue of Mollusca of Little Gull Island, Suffolk Co., New York." *Annals of the Lyceum of Natural History of New York,* vol. 8, p. 194–195.

Smith, Sanderson. 1865. "Notice of a Post-Pleiocene [sic] deposit on Gardiner's Island, Suffolk Co., N. Y." *Annals of the Lyceum of Natural History of New York,* vol. 8, p. 149–151.

Smith, Sanderson and Temple Prime. 1870. "Report on the Mollusca of Long Island, New York and of its dependencies." *Annals of the Lyceum of Natural History of New York,* vol. 9, p. 377–407 (most complete list published).

Smith, Sanderson. 1887. "Catalogue of the Mollusca of Staten Island." *Proceedings of the Natural Science Association, Staten Island,* Extra #5.

Prime, Henry. 1894. "Catalogue of land shells of Long Island, N.Y." *Nautilus,* vol. 8, p. 69, 70.

Balch, Francis Noyes. 1899. "List of Marine Mollusca of Cold Spring Harbor, Long Island. . . ." *Boston Society of Natural History,* vol. 28, p. 133–162, pl. 1.

Wheat, S. C. 1907. "List of Long Island Shells." *Bulletin of the Brooklyn Conchological Club,* vol. 1, p. 7–10.

Weeks, William H , Jr. 1908. "A Collecting Trip at Northport, N. Y." *Nautilus,* vol. 21, p. 98, 99.

Humphreys, Edwin W. 1909. "Recent fresh-water fossils [sic] from Bronx Borough, New York City." *Nautilus,* vol. 23, p. 10–11.

Jacot, Arthur. 1919. "Some Marine Mollusca About New York City." *Nautilus,* vol. 32, p. 90–94.

Jacot, Arthur. 1920. "On the Marine Mollusca of Staten Island, N.Y." *Nautilus,* vol. 33, p. 111–115.

Jacot, Arthur. 1921. "Notes on Marine Mollusca about New York City." *Nautilus,* vol. 34, p. 59–60.

Jacobson, Morris K. 1943. "Marine Mollusca of New York City." *Nautilus,* vol. 56, p. 139–144.

Abbott, R. Tucker. 1968. *A Guide to Field Identification, Seashells of North America.* Golden Press, New York, 280 pages, illustrations. This handy and informative identification guide is especially noteworthy because of its numerous excellent color illustrations. Figured are many species found in the New York City area.

Burch, John B. 1962. *How To Know The Eastern Land Snails. Pictured-Keys for determining the Land Snails of the United States occurring east of the Rocky Mountain Divide.* Wm. C. Brown Co., Dubuque, Iowa, 214 pages, 519 text figures. This book is printed in the form of an identification guide. Nearly all the species are superbly illustrated. All our local land shells, of course, find a place here too. This is a very useful guide.

The following articles deal in some way with the local molluscan fauna:

Clarke, Arthur H. (editor). 1970. "Papers on the Rare and Endangered Mollusks of North America." *Malacologia,* vol. 10, no. 1, p. 1–56. These reports include articles on eastern fresh-water, terrestrial and marine mollusks.

Jacobson, Morris K. 1965. "New Records for New York and New Jersey." *Nautilus,* vol. 78, no. 3, p. 83–86. Six gastropods, two each of marine, fresh-water and land, are recorded.

Morris, Percy A. 1963. "Some New York Sea Shells." *The Conservationist,* State of New York Conservation Department, Albany, vol. 17, no. 3, p. 20–27, 37, illustrations. Forty-seven of the more common marine gastropods and bivalves of the State are discussed and illustrated in color.

In addition, a large number of articles written by members of the New York Shell Club have appeared in the *New York Shell Club Notes,* a mimeographed periodical which appears ten times a year. Information regarding this periodical and the New York Shell Club may be obtained from The American Museum of Natural History, New York, New York.

NOTE: *The 15 additional species in the Supplement beginning on p. 137 are not listed in this Index.*

Index

Acmaea alveus (*see* A. testudinalis); testudinalis, *44*, 116, 127
Acteocina canaliculata (*see* Retusa)
Acteon punctostriatus, *70*, 129
Aequipecten irradians (*see* Pecten)
Amnicola limosa, *37*, 117, 127
Anachis avara, *56*, 58, 113, 129; avara similis, *57*
Anadara ovalis (*see* A. pexata); pexata, *74*, 112, 116, 133; transversa, *75*, 112, 116, 133
Anguispira alternata, *9*, 17, 117, 131; alternata fergusoni, *10*
Anodonta cataracta, *38*, 39, 133; implicata, *38*, 133
Anomia aculeata, *80*, 133; simplex, *79*, 106, 112, 116, 133
Aplexa hypnorum, *32*, 129
Arca pexata, transversa (*see* Anadara)
Arctica islandica, *85*, 112, 133
Astarte castanea, *86*, 112, 133; castanea picea, *86;* undata, *86*, 112, 133
Barnea costata (*see* Cyrtopleura); truncata, *101*, 102, 112, 115, 135
Bittium alternatum, *66*, 106, 115, 116, 127
Brachidontes demissus plicatulus, *82*, 112, 114, 115, 116, 133
Buccinum undatum, *60*, 113, 129
Bulimus tentaculatus, *36*, 127
Busycon canaliculatum, *62*, 112, 113, 114, 116, 129; carica, *63*, 113, 116, 129
Campeloma decisum, *36*, 127
Carychium exiguum, *20*, 129; exile, *21*, 129
Cepaea nemoralis, 9, *11*, 131
Cerithiopsis greenii, *66*, 113, 127
Chaetopleura apiculata, *103*, 127
Cionella lubrica, *18*, 117, 133
Cochlicopa lubrica (*see* Cionella)
Columella edentula, *22*, 131
Colus islandicus, *61*, 113, 129; stimpsoni, *62*, 129; stonei, *62*, 129
Congeria leucophaeata, *83*, 135
Corbula contracta, *98*, 112, 135
Crassostrea virginica, *76*, 106, 112, 114, 115, 116, 133
Crepidula convexa, *50*, 112, 113, 114, 115, 116, 127; fornicata, *49*, 50, 112, 113, 114, 115, 116, 127; plana, *50*, 112, 113, 115, 116, 127

Crucibulum striatum, *51*, 113, 127
Cylichna oryza, *70*, 113, 129
Cyrtopleura costata, *100*, 112, 115, 135
Discus cronkhitei catskillensis, *15*, 16, 117, 131; cronkhitei cronkhitei, *16*; rotundatus, *16*, 131
Divaricella quadrisulcata, *87*, 112, 135
Donax fossor, *93*, 97, 112, 113, 135; variabilis (*see* D. fossor)
Elliptio complanatus, *39*, 41, 133
Ensis directus, *94*, 112, 113, 114, 115, 116, 135
Eobania vermiculata, *107*
Epitonium angulatum, *46*, 127; humphreysii, *46*, 113, 115, 116, 127; lineata (*see* E. rupicola); multistriatum, *46*, 113, 127; rupicola, *45*, 115, 127
Euconulus chersinus, *22*, 131; fulvus, *22*, 23, 131
Eupleura caudata, *55*, 113, 114, 115, 127
Ferrissia fusca, *32*, 129; parallela, *33*, 129
Fossaria humilis modicella (*see* Lymnaea)
Gastrocopta contracta, *22*, 117, 131; tappaniana, *22*
Gemma gemma, *90*, 113, 116, 135; gemma manhattensis, *90*, 114, 115, 116; purpurea, *90*, 135
Goniobasis virginica, *34*, 127
Gyraulus parvus, *30*, 129
Haminoea solitaria, *64*, 114, 115, 129
Haplotrema concavum, *8*, 131
Hawaiia minuscula, *24*, 117, 131
Helicodiscus parallelus, *17*, 117, 131
Helisoma anceps, *29*, 30, 117, 129; trivolvis, *28*, 30, 129
Helix aperta, *109;* aspersa, *108;* aspersa maxima, *108;* melanostoma, *107;* pomatia, *107*
Hiatella arctica, *98*, 113, 135
Hydrobia minuta, *68*, 114, 115, 127
Lacuna vincta, *54*, 113, 127
Laevicardium mortoni, *88*, 113, 114, 115, 116, 135
Littorina irrorata, *52*, 127; littorea, *52*, 106, 112, 114, 115, 127; obtusata, *53*, 114, 116, 127; saxatilis, *53*, 114, 115, 116, 127
Lunatia heros (*see* Polinices)

Lymnaea auricularia, 26, 129; columella, 26, 129; humilis, 27, 129; palustris elodes, 26, 27, 117, 129
Lyonsia hyalina, 84, 135
Macoma balthica, 92, 114, 115, 135; calcarea, 92, 135
Mangelia cerina, 69, 113, 129; plicosa, 69, 113, 115, 116, 129
Margaritana margaritifera, 40, 133
Melampus bidentatus, 21, 64, 65, 114, 115, 129
Mercenaria mercenaria (see Venus)
Mesodesma arctatum, 97, 135
Mesodon thyroidus, 5, 11, 117, 131
Mitrella lunata, 57, 113, 129
Modiolus demissus plicatulus (see Brachidontes); modiolus, 82, 133
Mulinia lateralis, 96, 112, 113, 114, 115, 116, 135
Musculium species, 42, 117, 133
Mya arenaria, 99, 112, 114, 115, 116, 135
Mytilus edulis, 80, 83, 106, 112, 113, 114, 115, 116, 133
Nassarius obsoletus, 58, 112, 113, 114, 115, 116, 129; trivittatus, 59, 112, 113, 115, 116, 129; vibex, 60, 113, 115, 117, 129
Natica heros (see Polinices)
Noetia ponderosa, 76, 113, 133
Nucula proxima, 73, 114, 116, 133
Odostomia species, 67, 106, 114, 115, 129
Ostrea virginica (see Crassostrea)
Otala lactea, 106, 107
Ovatella myosotis, 21, 65, 114, 129
Oxychilus alliarius, 15, 131; cellarius, 14, 131; draparnaldi, 15, 117, 131
Pandora gouldiana, 83, 112, 135
Pecten grandis (see P. magellanicus); irradians, 77, 113, 114, 116, 133; magellanicus, 78, 113, 133
Periploma leanum, 84, 135
Petricola pholadiformis, 100, 112, 113, 114, 115, 116, 135; pholadiformis lata, 101
Physa ancillaria, 32, 129; heterostropha, 31, 117, 129
Phytia myosotis (see Ovatella)
Pisidium species, 42, 117, 133
Pitar morrhuana, 89, 113, 135
Placopecten magellanicus (see Pecten)
Planorbis antrosus; bicarinatus (see Helisoma anceps)
Planorbula jenksii, 29, 117, 129
Polinices duplicatus, 48, 112, 113, 114, 115, 117, 127; heros, 46, 48, 112, 113, 114, 115, 117, 127; triseriatus, 48, 113, 117, 127
Promenetus hudsonicus, 30, 117, 129
Punctum minutissimum, 24, 131
Retinella electrina, 14, 117, 131; indentata, 14, 117, 131
Retusa canaliculata, 69, 114, 115, 129
Saxicava arctica (see Hiatella)
Seila adamsi, 66, 114, 127
Siliqua costata, 94, 113, 135
Solemya velum, 73, 114, 116, 133
Sphaerium species, 42, 117, 133
Spirula spirula, 70, 135
Spisula solidissima, 96, 112, 113, 116, 135
Stenotrema fraternum, 7, 131; hirsutum, 7, 117, 131
Strobilops aenea, 23, 131; affinis, 23, 131; labyrinthica, 23, 131
Succinea avara, 10, 26, 133; ovalis, 10, 26, 117, 133
Tagelus divisus, 94, 135; gibbus (see T. plebeius); plebeius, 94, 113, 114, 135
Tellina agilis, 92, 112, 113, 114, 115, 116, 135; tenera (see T. agilis)
Teredo navalis, 102, 113, 135
Thais lapillus, 56, 127
Theba pisana, 108
Triodopsis albolabris, 3, 6, 117, 131; albolabris traversensis, 4, 131; denotata, 4, 131; tridentata, 4, 117, 131
Triphora perversa nigrocincta, 67, 114, 127
Turbonilla species, 68, 114, 129
Urosalpinx cinerea, 54, 56, 114, 115, 127; cinerea atkinae, 55
Vallonia costata, 20, 133; excentrica, 20, 133; pulchella, 19, 24, 117, 133
Valvata tricarinata, 37, 117, 127
Venericardia borealis, 87, 113, 133
Ventridens ligera, 6, 131
Venus mercenaria, 85, 88, 90, 106, 113, 114, 115, 116, 135; mercenaria alba, 89; mercenaria notata, 89, 106, 115
Vertigo gouldii, 22, 131; ovata, 22, 131
Viviparus contectoides, 35, 127; malleatus, 34, 35, 127
Volsella demissa (see Brachidontes); modiolus (see Modiolus)
Yoldia limatula, 74, 133; sapotilla, 74, 133
Zirfaea crispata, 102, 113, 135
Zonitoides arboreus, 12, 13, 14, 16, 117, 131; nitidus, 13, 117, 131

A CATALOGUE OF SELECTED DOVER BOOKS
IN ALL FIELDS OF INTEREST

A CATALOGUE OF SELECTED DOVER BOOKS
IN ALL FIELDS OF INTEREST

AMERICA'S OLD MASTERS, James T. Flexner. Four men emerged unexpectedly from provincial 18th century America to leadership in European art: Benjamin West, J. S. Copley, C. R. Peale, Gilbert Stuart. Brilliant coverage of lives and contributions. Revised, 1967 edition. 69 plates. 365pp. of text.
21806-6 Paperbound $3.00

FIRST FLOWERS OF OUR WILDERNESS: AMERICAN PAINTING, THE COLONIAL PERIOD, James T. Flexner. Painters, and regional painting traditions from earliest Colonial times up to the emergence of Copley, West and Peale Sr., Foster, Gustavus Hesselius, Feke, John Smibert and many anonymous painters in the primitive manner. Engaging presentation, with 162 illustrations. xxii + 368pp.
22180-6 Paperbound $3.50

THE LIGHT OF DISTANT SKIES: AMERICAN PAINTING, 1760-1835, James T. Flexner. The great generation of early American painters goes to Europe to learn and to teach: West, Copley, Gilbert Stuart and others. Allston, Trumbull, Morse; also contemporary American painters—primitives, derivatives, academics—who remained in America. 102 illustrations. xiii + 306pp.
22179-2 Paperbound $3.00

A HISTORY OF THE RISE AND PROGRESS OF THE ARTS OF DESIGN IN THE UNITED STATES, William Dunlap. Much the richest mine of information on early American painters, sculptors, architects, engravers, miniaturists, etc. The only source of information for scores of artists, the major primary source for many others. Unabridged reprint of rare original 1834 edition, with new introduction by James T. Flexner, and 394 new illustrations. Edited by Rita Weiss. $6\frac{5}{8}$ x $9\frac{5}{8}$.
21695-0, 21696-9, 21697-7 Three volumes, Paperbound $13.50

EPOCHS OF CHINESE AND JAPANESE ART, Ernest F. Fenollosa. From primitive Chinese art to the 20th century, thorough history, explanation of every important art period and form, including Japanese woodcuts; main stress on China and Japan, but Tibet, Korea also included. Still unexcelled for its detailed, rich coverage of cultural background, aesthetic elements, diffusion studies, particularly of the historical period. 2nd, 1913 edition. 242 illustrations. lii + 439pp. of text.
20364-6, 20365-4 Two volumes, Paperbound $6.00

THE GENTLE ART OF MAKING ENEMIES, James A. M. Whistler. Greatest wit of his day deflates Oscar Wilde, Ruskin, Swinburne; strikes back at inane critics, exhibitions, art journalism; aesthetics of impressionist revolution in most striking form. Highly readable classic by great painter. Reproduction of edition designed by Whistler. Introduction by Alfred Werner. xxxvi + 334pp.
21875-9 Paperbound $2.50

CATALOGUE OF DOVER BOOKS

ALPHABETS AND ORNAMENTS, Ernst Lehner. Well-known pictorial source for decorative alphabets, script examples, cartouches, frames, decorative title pages, calligraphic initials, borders, similar material. 14th to 19th century, mostly European. Useful in almost any graphic arts designing, varied styles. 750 illustrations. 256pp. 7 x 10. 21905-4 Paperbound $4.00

PAINTING: A CREATIVE APPROACH, Norman Colquhoun. For the beginner simple guide provides an instructive approach to painting: major stumbling blocks for beginner; overcoming them, technical points; paints and pigments; oil painting; watercolor and other media and color. New section on "plastic" paints. Glossary. Formerly *Paint Your Own Pictures.* 221pp. 22000-1 Paperbound $1.75

THE ENJOYMENT AND USE OF COLOR, Walter Sargent. Explanation of the relations between colors themselves and between colors in nature and art, including hundreds of little-known facts about color values, intensities, effects of high and low illumination, complementary colors. Many practical hints for painters, references to great masters. 7 color plates, 29 illustrations. x + 274pp.
20944-X Paperbound $2.50

THE NOTEBOOKS OF LEONARDO DA VINCI, compiled and edited by Jean Paul Richter. 1566 extracts from original manuscripts reveal the full range of Leonardo's versatile genius: all his writings on painting, sculpture, architecture, anatomy, astronomy, geography, topography, physiology, mining, music, etc., in both Italian and English, with 186 plates of manuscript pages and more than 500 additional drawings. Includes studies for the Last Supper, the lost Sforza monument, and other works. Total of xlvii + 866pp. $7\frac{7}{8}$ x $10\frac{3}{4}$.
22572-0, 22573-9 Two volumes, Paperbound $10.00

MONTGOMERY WARD CATALOGUE OF 1895. Tea gowns, yards of flannel and pillow-case lace, stereoscopes, books of gospel hymns, the New Improved Singer Sewing Machine, side saddles, milk skimmers, straight-edged razors, high-button shoes, spittoons, and on and on . . . listing some 25,000 items, practically all illustrated. Essential to the shoppers of the 1890's, it is our truest record of the spirit of the period. Unaltered reprint of Issue No. 57, Spring and Summer 1895. Introduction by Boris Emmet. Innumerable illustrations. xiii + 624pp. $8\frac{1}{2}$ x $11\frac{5}{8}$.
22377-9 Paperbound $6.95

THE CRYSTAL PALACE EXHIBITION ILLUSTRATED CATALOGUE (LONDON, 1851). One of the wonders of the modern world—the Crystal Palace Exhibition in which all the nations of the civilized world exhibited their achievements in the arts and sciences—presented in an equally important illustrated catalogue. More than 1700 items pictured with accompanying text—ceramics, textiles, cast-iron work, carpets, pianos, sleds, razors, wall-papers, billiard tables, beehives, silverware and hundreds of other artifacts—represent the focal point of Victorian culture in the Western World. Probably the largest collection of Victorian decorative art ever assembled—indispensable for antiquarians and designers. Unabridged republication of the Art-Journal Catalogue of the Great Exhibition of 1851, with all terminal essays. New introduction by John Gloag, F.S.A. xxxiv + 426pp. 9 x 12.
22503-8 Paperbound $4.50

CATALOGUE OF DOVER BOOKS

THE ARCHITECTURE OF COUNTRY HOUSES, Andrew J. Downing. Together with Vaux's *Villas and Cottages* this is the basic book for Hudson River Gothic architecture of the middle Victorian period. Full, sound discussions of general aspects of housing, architecture, style, decoration, furnishing, together with scores of detailed house plans, illustrations of specific buildings, accompanied by full text. Perhaps the most influential single American architectural book. 1850 edition. Introduction by J. Stewart Johnson. 321 figures, 34 architectural designs. xvi + 560pp.
22003-6 Paperbound $4.00

LOST EXAMPLES OF COLONIAL ARCHITECTURE, John Mead Howells. Full-page photographs of buildings that have disappeared or been so altered as to be denatured, including many designed by major early American architects. 245 plates. xvii + 248pp. 7⅞ x 10¾.
21143-6 Paperbound $3.00

DOMESTIC ARCHITECTURE OF THE AMERICAN COLONIES AND OF THE EARLY REPUBLIC, Fiske Kimball. Foremost architect and restorer of Williamsburg and Monticello covers nearly 200 homes between 1620-1825. Architectural details, construction, style features, special fixtures, floor plans, etc. Generally considered finest work in its area. 219 illustrations of houses, doorways, windows, capital mantels. xx + 314pp. 7⅞ x 10¾.
21743-4 Paperbound $3.50

EARLY AMERICAN ROOMS: 1650-1858, edited by Russell Hawes Kettell. Tour of 12 rooms, each representative of a different era in American history and each furnished, decorated, designed and occupied in the style of the era. 72 plans and elevations, 8-page color section, etc., show fabrics, wall papers, arrangements, etc. Full descriptive text. xvii + 200pp. of text. 8⅜ x 11¼.
21633-0 Paperbound $5.00

THE FITZWILLIAM VIRGINAL BOOK, edited by J. Fuller Maitland and W. B. Squire. Full modern printing of famous early 17th-century ms. volume of 300 works by Morley, Byrd, Bull, Gibbons, etc. For piano or other modern keyboard instrument; easy to read format. xxxvi + 938pp. 8⅜ x 11.
21068-5, 21069-3 Two volumes, Paperbound $8.00

HARPSICHORD MUSIC, Johann Sebastian Bach. Bach Gesellschaft edition. A rich selection of Bach's masterpieces for the harpsichord: the six English Suites, six French Suites, the six Partitas (Clavierübung part I), the Goldberg Variations (Clavierübung part IV), the fifteen Two-Part Inventions and the fifteen Three-Part Sinfonias. Clearly reproduced on large sheets with ample margins; eminently playable. vi + 312pp. 8⅛ x 11.
22360-4 Paperbound $5.00

THE MUSIC OF BACH: AN INTRODUCTION, Charles Sanford Terry. A fine, nontechnical introduction to Bach's music, both instrumental and vocal. Covers organ music, chamber music, passion music, other types. Analyzes themes, developments, innovations. x + 114pp.
21075-8 Paperbound $1.25

BEETHOVEN AND HIS NINE SYMPHONIES, Sir George Grove. Noted British musicologist provides best history, analysis, commentary on symphonies. Very thorough, rigorously accurate; necessary to both advanced student and amateur music lover. 436 musical passages. vii + 407 pp.
20334-4 Paperbound $2.25

CATALOGUE OF DOVER BOOKS

JOHANN SEBASTIAN BACH, Philipp Spitta. One of the great classics of musicology, this definitive analysis of Bach's music (and life) has never been surpassed. Lucid, nontechnical analyses of hundreds of pieces (30 pages devoted to St. Matthew Passion, 26 to B Minor Mass). Also includes major analysis of 18th-century music. 450 musical examples. 40-page musical supplement. Total of xx + 1799pp.
(EUK) 22278-0, 22279-9 Two volumes, Clothbound $15.00

MOZART AND HIS PIANO CONCERTOS, Cuthbert Girdlestone. The only full-length study of an important area of Mozart's creativity. Provides detailed analyses of all 23 concertos, traces inspirational sources. 417 musical examples. Second edition. 509pp. (USO) 21271-8 Paperbound $3.50

THE PERFECT WAGNERITE: A COMMENTARY ON THE NIBLUNG'S RING, George Bernard Shaw. Brilliant and still relevant criticism in remarkable essays on Wagner's Ring cycle, Shaw's ideas on political and social ideology behind the plots, role of Leitmotifs, vocal requisites, etc. Prefaces. xxi + 136pp.
21707-8 Paperbound $1.50

DON GIOVANNI, W. A. Mozart. Complete libretto, modern English translation; biographies of composer and librettist; accounts of early performances and critical reaction. Lavishly illustrated. All the material you need to understand and appreciate this great work. Dover Opera Guide and Libretto Series; translated and introduced by Ellen Bleiler. 92 illustrations. 209pp.
21134-7 Paperbound $1.50

HIGH FIDELITY SYSTEMS: A LAYMAN'S GUIDE, Roy F. Allison. All the basic information you need for setting up your own audio system: high fidelity and stereo record players, tape records, F.M. Connections, adjusting tone arm, cartridge, checking needle alignment, positioning speakers, phasing speakers, adjusting hums, trouble-shooting, maintenance, and similar topics. Enlarged 1965 edition. More than 50 charts, diagrams, photos. iv + 91pp. 21514-8 Paperbound $1.25

REPRODUCTION OF SOUND, Edgar Villchur. Thorough coverage for laymen of high fidelity systems, reproducing systems in general, needles, amplifiers, preamps, loudspeakers, feedback, explaining physical background. "A rare talent for making technicalities vividly comprehensible," R. Darrell, *High Fidelity*. 69 figures. iv + 92pp. 21515-6 Paperbound $1.00

HEAR ME TALKIN' TO YA: THE STORY OF JAZZ AS TOLD BY THE MEN WHO MADE IT, Nat Shapiro and Nat Hentoff. Louis Armstrong, Fats Waller, Jo Jones, Clarence Williams, Billy Holiday, Duke Ellington, Jelly Roll Morton and dozens of other jazz greats tell how it was in Chicago's South Side, New Orleans, depression Harlem and the modern West Coast as jazz was born and grew. xvi + 429pp.
21726-4 Paperbound $2.50

FABLES OF AESOP, translated by Sir Roger L'Estrange. A reproduction of the very rare 1931 Paris edition; a selection of the most interesting fables, together with 50 imaginative drawings by Alexander Calder. v + 128pp. 6½x9¼.
21780-9 Paperbound $1.25

CATALOGUE OF DOVER BOOKS

POEMS OF ANNE BRADSTREET, edited with an introduction by Robert Hutchinson. A new selection of poems by America's first poet and perhaps the first significant woman poet in the English language. 48 poems display her development in works of considerable variety—love poems, domestic poems, religious meditations, formal elegies, "quaternions," etc. Notes, bibliography. viii + 222pp.
22160-1 Paperbound $2.00

THREE GOTHIC NOVELS: THE CASTLE OF OTRANTO BY HORACE WALPOLE; VATHEK BY WILLIAM BECKFORD; THE VAMPYRE BY JOHN POLIDORI, WITH FRAGMENT OF A NOVEL BY LORD BYRON, edited by E. F. Bleiler. The first Gothic novel, by Walpole; the finest Oriental tale in English, by Beckford; powerful Romantic supernatural story in versions by Polidori and Byron. All extremely important in history of literature; all still exciting, packed with supernatural thrills, ghosts, haunted castles, magic, etc. xl + 291pp.
21232-7 Paperbound $2.00

THE BEST TALES OF HOFFMANN, E. T. A. Hoffmann. 10 of Hoffmann's most important stories, in modern re-editings of standard translations: Nutcracker and the King of Mice, Signor Formica, Automata, The Sandman, Rath Krespel, The Golden Flowerpot, Master Martin the Cooper, The Mines of Falun, The King's Betrothed, A New Year's Eve Adventure. 7 illustrations by Hoffmann. Edited by E. F. Bleiler. xxxix + 419pp.
21793-0 Paperbound $2.50

GHOST AND HORROR STORIES OF AMBROSE BIERCE, Ambrose Bierce. 23 strikingly modern stories of the horrors latent in the human mind: The Eyes of the Panther, The Damned Thing, An Occurrence at Owl Creek Bridge, An Inhabitant of Carcosa, etc., plus the dream-essay, Visions of the Night. Edited by E. F. Bleiler. xxii + 199pp.
20767-6 Paperbound $1.50

BEST GHOST STORIES OF J. S. LEFANU, J. Sheridan LeFanu. Finest stories by Victorian master often considered greatest supernatural writer of all. Carmilla, Green Tea, The Haunted Baronet, The Familiar, and 12 others. Most never before available in the U. S. A. Edited by E. F. Bleiler. 8 illustrations from Victorian publications. xvii + 467pp.
20415-4 Paperbound $2.50

THE TIME STREAM, THE GREATEST ADVENTURE, AND THE PURPLE SAPPHIRE—THREE SCIENCE FICTION NOVELS, John Taine (Eric Temple Bell). Great American mathematician was also foremost science fiction novelist of the 1920's. *The Time Stream,* one of all-time classics, uses concepts of circular time; *The Greatest Adventure,* incredibly ancient biological experiments from Antarctica threaten to escape; *The Purple Sapphire,* superscience, lost races in Central Tibet, survivors of the Great Race. 4 illustrations by Frank R. Paul. v + 532pp.
21180-0 Paperbound $3.00

SEVEN SCIENCE FICTION NOVELS, H. G. Wells. The standard collection of the great novels. Complete, unabridged. *First Men in the Moon, Island of Dr. Moreau, War of the Worlds, Food of the Gods, Invisible Man, Time Machine, In the Days of the Comet.* Not only science fiction fans, but every educated person owes it to himself to read these novels. 1015pp.
20264-X Clothbound $5.00

CATALOGUE OF DOVER BOOKS

EAST O' THE SUN AND WEST O' THE MOON, George W. Dasent. Considered the best of all translations of these Norwegian folk tales, this collection has been enjoyed by generations of children (and folklorists too). Includes True and Untrue, Why the Sea is Salt, East O' the Sun and West O' the Moon, Why the Bear is Stumpy-Tailed, Boots and the Troll, The Cock and the Hen, Rich Peter the Pedlar, and 52 more. The only edition with all 59 tales. 77 illustrations by Erik Werenskiold and Theodor Kittelsen. xv + 418pp. 22521-6 Paperbound $3.00

GOOPS AND HOW TO BE THEM, Gelett Burgess. Classic of tongue-in-cheek humor, masquerading as etiquette book. 87 verses, twice as many cartoons, show mischievous Goops as they demonstrate to children virtues of table manners, neatness, courtesy, etc. Favorite for generations. viii + 88pp. 6½ x 9¼. 22233-0 Paperbound $1.25

ALICE'S ADVENTURES UNDER GROUND, Lewis Carroll. The first version, quite different from the final *Alice in Wonderland,* printed out by Carroll himself with his own illustrations. Complete facsimile of the "million dollar" manuscript Carroll gave to Alice Liddell in 1864. Introduction by Martin Gardner. viii + 96pp. Title and dedication pages in color. 21482-6 Paperbound $1.00

THE BROWNIES, THEIR BOOK, Palmer Cox. Small as mice, cunning as foxes, exuberant and full of mischief, the Brownies go to the zoo, toy shop, seashore, circus, etc., in 24 verse adventures and 266 illustrations. Long a favorite, since their first appearance in St. Nicholas Magazine. xi + 144pp. 6⅝ x 9¼. 21265-3 Paperbound $1.50

SONGS OF CHILDHOOD, Walter De La Mare. Published (under the pseudonym Walter Ramal) when De La Mare was only 29, this charming collection has long been a favorite children's book. A facsimile of the first edition in paper, the 47 poems capture the simplicity of the nursery rhyme and the ballad, including such lyrics as I Met Eve, Tartary, The Silver Penny. vii + 106pp. 21972-0 Paperbound $1.25

THE COMPLETE NONSENSE OF EDWARD LEAR, Edward Lear. The finest 19th-century humorist-cartoonist in full: all nonsense limericks, zany alphabets, Owl and Pussycat, songs, nonsense botany, and more than 500 illustrations by Lear himself. Edited by Holbrook Jackson. xxix + 287pp. (USO) 20167-8 Paperbound $1.75

BILLY WHISKERS: THE AUTOBIOGRAPHY OF A GOAT, Frances Trego Montgomery. A favorite of children since the early 20th century, here are the escapades of that rambunctious, irresistible and mischievous goat—Billy Whiskers. Much in the spirit of *Peck's Bad Boy,* this is a book that children never tire of reading or hearing. All the original familiar illustrations by W. H. Fry are included: 6 color plates, 18 black and white drawings. 159pp. 22345-0 Paperbound $2.00

MOTHER GOOSE MELODIES. Faithful republication of the fabulously rare Munroe and Francis "copyright 1833" Boston edition—the most important Mother Goose collection, usually referred to as the "original." Familiar rhymes plus many rare ones, with wonderful old woodcut illustrations. Edited by E. F. Bleiler. 128pp. 4½ x 6⅜. 22577-1 Paperbound $1.25

CATALOGUE OF DOVER BOOKS

How to Know the Wild Flowers, Mrs. William Starr Dana. This is the classical book of American wildflowers (of the Eastern and Central United States), used by hundreds of thousands. Covers over 500 species, arranged in extremely easy to use color and season groups. Full descriptions, much plant lore. This Dover edition is the fullest ever compiled, with tables of nomenclature changes. 174 full-page plates by M. Satterlee. xii + 418pp. 20332-8 Paperbound $2.50

Our Plant Friends and Foes, William Atherton DuPuy. History, economic importance, essential botanical information and peculiarities of 25 common forms of plant life are provided in this book in an entertaining and charming style. Covers food plants (potatoes, apples, beans, wheat, almonds, bananas, etc.), flowers (lily, tulip, etc.), trees (pine, oak, elm, etc.), weeds, poisonous mushrooms and vines, gourds, citrus fruits, cotton, the cactus family, and much more. 108 illustrations. xiv + 290pp. 22272-1 Paperbound $2.00

How to Know the Ferns, Frances T. Parsons. Classic survey of Eastern and Central ferns, arranged according to clear, simple identification key. Excellent introduction to greatly neglected nature area. 57 illustrations and 42 plates. xvi + 215pp. 20740-4 Paperbound $1.75

Manual of the Trees of North America, Charles S. Sargent. America's foremost dendrologist provides the definitive coverage of North American trees and tree-like shrubs. 717 species fully described and illustrated: exact distribution, down to township; full botanical description; economic importance; description of sub-species and races; habitat, growth data; similar material. Necessary to every serious student of tree-life. Nomenclature revised to present. Over 100 locating keys. 783 illustrations. lii + 934pp. 20277-1, 20278-X Two volumes, Paperbound $6.00

Our Northern Shrubs, Harriet L. Keeler. Fine non-technical reference work identifying more than 225 important shrubs of Eastern and Central United States and Canada. Full text covering botanical description, habitat, plant lore, is paralleled with 205 full-page photographs of flowering or fruiting plants. Nomenclature revised by Edward G. Voss. One of few works concerned with shrubs. 205 plates, 35 drawings. xxviii + 521pp. 21989-5 Paperbound $3.75

The Mushroom Handbook, Louis C. C. Krieger. Still the best popular handbook: full descriptions of 259 species, cross references to another 200. Extremely thorough text enables you to identify, know all about any mushroom you are likely to meet in eastern and central U. S. A.: habitat, luminescence, poisonous qualities, use, folklore, etc. 32 color plates show over 50 mushrooms, also 126 other illustrations. Finding keys. vii + 560pp. 21861-9 Paperbound $3.95

Handbook of Birds of Eastern North America, Frank M. Chapman. Still much the best single-volume guide to the birds of Eastern and Central United States. Very full coverage of 675 species, with descriptions, life habits, distribution, similar data. All descriptions keyed to two-page color chart. With this single volume the average birdwatcher needs no other books. 1931 revised edition. 195 illustrations. xxxvi + 581pp. 21489-3 Paperbound $3.25

CATALOGUE OF DOVER BOOKS

"ESSENTIAL GRAMMAR" SERIES

All you really need to know about modern, colloquial grammar. Many educational shortcuts help you learn faster, understand better. Detailed cognate lists teach you to recognize similarities between English and foreign words and roots—make learning vocabulary easy and interesting. Excellent for independent study or as a supplement to record courses.

ESSENTIAL FRENCH GRAMMAR, Seymour Resnick. 2500-item cognate list. 159pp.
(EBE) 20419-7 Paperbound $1.25

ESSENTIAL GERMAN GRAMMAR, Guy Stern and Everett F. Bleiler. Unusual shortcuts on noun declension, word order, compound verbs. 124pp.
(EBE) 20422-7 Paperbound $1.25

ESSENTIAL ITALIAN GRAMMAR, Olga Ragusa. 111pp.
(EBE) 20779-X Paperbound $1.25

ESSENTIAL JAPANESE GRAMMAR, Everett F. Bleiler. In Romaji transcription; no characters needed. Japanese grammar is regular and simple. 156pp.
21027-8 Paperbound $1.25

ESSENTIAL PORTUGUESE GRAMMAR, Alexander da R. Prista. vi + 114pp.
21650-0 Paperbound $1.25

ESSENTIAL SPANISH GRAMMAR, Seymour Resnick. 2500 word cognate list. 115pp.
(EBE) 20780-3 Paperbound $1.25

ESSENTIAL ENGLISH GRAMMAR, Philip Gucker. Combines best features of modern, functional and traditional approaches. For refresher, class use, home study. x + 177pp.
21649-7 Paperbound $1.25

A PHRASE AND SENTENCE DICTIONARY OF SPOKEN SPANISH. Prepared for U. S. War Department by U. S. linguists. As above, unit is idiom, phrase or sentence rather than word. English-Spanish and Spanish-English sections contain modern equivalents of over 18,000 sentences. Introduction and appendix as above. iv + 513pp.
20495-2 Paperbound $2.00

A PHRASE AND SENTENCE DICTIONARY OF SPOKEN RUSSIAN. Dictionary prepared for U. S. War Department by U. S. linguists. Basic unit is not the word, but the idiom, phrase or sentence. English-Russian and Russian-English sections contain modern equivalents for over 30,000 phrases. Grammatical introduction covers phonetics, writing, syntax. Appendix of word lists for food, numbers, geographical names, etc. vi + 573 pp. $6\frac{1}{8}$ x $9\frac{1}{4}$.
20496-0 Paperbound $3.00

CONVERSATIONAL CHINESE FOR BEGINNERS, Morris Swadesh. Phonetic system, beginner's course in Pai Hua Mandarin Chinese covering most important, most useful speech patterns. Emphasis on modern colloquial usage. Formerly *Chinese in Your Pocket*. xvi + 158pp.
21123-1 Paperbound $1.50

CATALOGUE OF DOVER BOOKS

LAST AND FIRST MEN AND STAR MAKER, TWO SCIENCE FICTION NOVELS, Olaf Stapledon. Greatest future histories in science fiction. In the first, human intelligence is the "hero," through strange paths of evolution, interplanetary invasions, incredible technologies, near extinctions and reemergences. Star Maker describes the quest of a band of star rovers for intelligence itself, through time and space: weird inhuman civilizations, crustacean minds, symbiotic worlds, etc. Complete, unabridged. v + 438pp. 21962-3 Paperbound $2.00

THREE PROPHETIC NOVELS, H. G. WELLS. Stages of a consistently planned future for mankind. *When the Sleeper Wakes,* and *A Story of the Days to Come,* anticipate *Brave New World* and *1984,* in the 21st Century; *The Time Machine,* only complete version in print, shows farther future and the end of mankind. All show Wells's greatest gifts as storyteller and novelist. Edited by E. F. Bleiler. x + 335pp. (USO) 20605-X Paperbound $2.00

THE DEVIL'S DICTIONARY, Ambrose Bierce. America's own Oscar Wilde—Ambrose Bierce—offers his barbed iconoclastic wisdom in over 1,000 definitions hailed by H. L. Mencken as "some of the most gorgeous witticisms in the English language." 145pp. 20487-1 Paperbound $1.25

MAX AND MORITZ, Wilhelm Busch. Great children's classic, father of comic strip, of two bad boys, Max and Moritz. Also Ker and Plunk (Plisch und Plumm), Cat and Mouse, Deceitful Henry, Ice-Peter, The Boy and the Pipe, and five other pieces. Original German, with English translation. Edited by H. Arthur Klein; translations by various hands and H. Arthur Klein. vi + 216pp.
20181-3 Paperbound $1.50

PIGS IS PIGS AND OTHER FAVORITES, Ellis Parker Butler. The title story is one of the best humor short stories, as Mike Flannery obfuscates biology and English. Also included, That Pup of Murchison's, The Great American Pie Company, and Perkins of Portland. 14 illustrations. v + 109pp. 21532-6 Paperbound $1.00

THE PETERKIN PAPERS, Lucretia P. Hale. It takes genius to be as stupidly mad as the Peterkins, as they decide to become wise, celebrate the "Fourth," keep a cow, and otherwise strain the resources of the Lady from Philadelphia. Basic book of American humor. 153 illustrations. 219pp. 20794-3 Paperbound $1.25

PERRAULT'S FAIRY TALES, translated by A. E. Johnson and S. R. Littlewood, with 34 full-page illustrations by Gustave Doré. All the original Perrault stories—Cinderella, Sleeping Beauty, Bluebeard, Little Red Riding Hood, Puss in Boots, Tom Thumb, etc.—with their witty verse morals and the magnificent illustrations of Doré. One of the five or six great books of European fairy tales. viii + 117pp. 8⅛ x 11. 22311-6 Paperbound $2.00

OLD HUNGARIAN FAIRY TALES, Baroness Orczy. Favorites translated and adapted by author of the *Scarlet Pimpernel.* Eight fairy tales include "The Suitors of Princess Fire-Fly," "The Twin Hunchbacks," "Mr. Cuttlefish's Love Story," and "The Enchanted Cat." This little volume of magic and adventure will captivate children as it has for generations. 90 drawings by Montagu Barstow. 96pp.
(USO) 22293-4 Paperbound $1.95

CATALOGUE OF DOVER BOOKS

AMERICAN FOOD AND GAME FISHES, David S. Jordan and Barton W. Evermann. Definitive source of information, detailed and accurate enough to enable the sportsman and nature lover to identify conclusively some 1,000 species and sub-species of North American fish, sought for food or sport. Coverage of range, physiology, habits, life history, food value. Best methods of capture, interest to the angler, advice on bait, fly-fishing, etc. 338 drawings and photographs. 1 + 574pp. $6\frac{5}{8}$ x $9\frac{3}{8}$.
22383-1 Paperbound $4.50

THE FROG BOOK, Mary C. Dickerson. Complete with extensive finding keys, over 300 photographs, and an introduction to the general biology of frogs and toads, this is the classic non-technical study of Northeastern and Central species. 58 species; 290 photographs and 16 color plates. xvii + 253pp.
21973-9 Paperbound $4.00

THE MOTH BOOK: A GUIDE TO THE MOTHS OF NORTH AMERICA, William J. Holland. Classical study, eagerly sought after and used for the past 60 years. Clear identification manual to more than 2,000 different moths, largest manual in existence. General information about moths, capturing, mounting, classifying, etc., followed by species by species descriptions. 263 illustrations plus 48 color plates show almost every species, full size. 1968 edition, preface, nomenclature changes by A. E. Brower. xxiv + 479pp. of text. $6\frac{1}{2}$ x $9\frac{1}{4}$.
21948-8 Paperbound $5.00

THE SEA-BEACH AT EBB-TIDE, Augusta Foote Arnold. Interested amateur can identify hundreds of marine plants and animals on coasts of North America; marine algae; seaweeds; squids; hermit crabs; horse shoe crabs; shrimps; corals; sea anemones; etc. Species descriptions cover: structure; food; reproductive cycle; size; shape; color; habitat; etc. Over 600 drawings. 85 plates. xii + 490pp.
21949-6 Paperbound $3.50

COMMON BIRD SONGS, Donald J. Borror. $33\frac{1}{3}$ 12-inch record presents songs of 60 important birds of the eastern United States. A thorough, serious record which provides several examples for each bird, showing different types of song, individual variations, etc. Inestimable identification aid for birdwatcher. 32-page booklet gives text about birds and songs, with illustration for each bird.
21829-5 Record, book, album. Monaural. $2.75

FADS AND FALLACIES IN THE NAME OF SCIENCE, Martin Gardner. Fair, witty appraisal of cranks and quacks of science: Atlantis, Lemuria, hollow earth, flat earth, Velikovsky, orgone energy, Dianetics, flying saucers, Bridey Murphy, food fads, medical fads, perpetual motion, etc. Formerly "In the Name of Science." x + 363pp.
20394-8 Paperbound $2.00

HOAXES, Curtis D. MacDougall. Exhaustive, unbelievably rich account of great hoaxes: Locke's moon hoax, Shakespearean forgeries, sea serpents, Loch Ness monster, Cardiff giant, John Wilkes Booth's mummy, Disumbrationist school of art, dozens more; also journalism, psychology of hoaxing. 54 illustrations. xi + 338pp.
20465-0 Paperbound $2.75

CATALOGUE OF DOVER BOOKS

THE PRINCIPLES OF PSYCHOLOGY, William James. The famous long course, complete and unabridged. Stream of thought, time perception, memory, experimental methods—these are only some of the concerns of a work that was years ahead of its time and still valid, interesting, useful. 94 figures. Total of xviii + 1391pp.
20381-6, 20382-4 Two volumes, Paperbound $6.00

THE STRANGE STORY OF THE QUANTUM, Banesh Hoffmann. Non-mathematical but thorough explanation of work of Planck, Einstein, Bohr, Pauli, de Broglie, Schrödinger, Heisenberg, Dirac, Feynman, etc. No technical background needed. "Of books attempting such an account, this is the best," Henry Margenau, Yale. 40-page "Postscript 1959." xii + 285pp. 20518-5 Paperbound $2.00

THE RISE OF THE NEW PHYSICS, A. d'Abro. Most thorough explanation in print of central core of mathematical physics, both classical and modern; from Newton to Dirac and Heisenberg. Both history and exposition; philosophy of science, causality, explanations of higher mathematics, analytical mechanics, electromagnetism, thermodynamics, phase rule, special and general relativity, matrices. No higher mathematics needed to follow exposition, though treatment is elementary to intermediate in level. Recommended to serious student who wishes verbal understanding. 97 illustrations. xvii + 982pp. 20003-5, 20004-3 Two volumes, Paperbound $5.50

GREAT IDEAS OF OPERATIONS RESEARCH, Jagjit Singh. Easily followed non-technical explanation of mathematical tools, aims, results: statistics, linear programming, game theory, queueing theory, Monte Carlo simulation, etc. Uses only elementary mathematics. Many case studies, several analyzed in detail. Clarity, breadth make this excellent for specialist in another field who wishes background. 41 figures. x + 228pp. 21886-4 Paperbound $2.25

GREAT IDEAS OF MODERN MATHEMATICS: THEIR NATURE AND USE, Jagjit Singh. Internationally famous expositor, winner of Unesco's Kalinga Award for science popularization explains verbally such topics as differential equations, matrices, groups, sets, transformations, mathematical logic and other important modern mathematics, as well as use in physics, astrophysics, and similar fields. Superb exposition for layman, scientist in other areas. viii + 312pp.
20587-8 Paperbound $2.25

GREAT IDEAS IN INFORMATION THEORY, LANGUAGE AND CYBERNETICS, Jagjit Singh. The analog and digital computers, how they work, how they are like and unlike the human brain, the men who developed them, their future applications, computer terminology. An essential book for today, even for readers with little math. Some mathematical demonstrations included for more advanced readers. 118 figures. Tables. ix + 338pp. 21694-2 Paperbound $2.25

CHANCE, LUCK AND STATISTICS, Horace C. Levinson. Non-mathematical presentation of fundamentals of probability theory and science of statistics and their applications. Games of chance, betting odds, misuse of statistics, normal and skew distributions, birth rates, stock speculation, insurance. Enlarged edition. Formerly "The Science of Chance." xiii + 357pp. 21007-3 Paperbound $2.00

CATALOGUE OF DOVER BOOKS

MATHEMATICAL PUZZLES FOR BEGINNERS AND ENTHUSIASTS, Geoffrey Mott-Smith. 189 puzzles from easy to difficult—involving arithmetic, logic, algebra, properties of digits, probability, etc.—for enjoyment and mental stimulus. Explanation of mathematical principles behind the puzzles. 135 illustrations. viii + 248pp.
20198-8 Paperbound $1.25

PAPER FOLDING FOR BEGINNERS, William D. Murray and Francis J. Rigney. Easiest book on the market, clearest instructions on making interesting, beautiful origami. Sail boats, cups, roosters, frogs that move legs, bonbon boxes, standing birds, etc. 40 projects; more than 275 diagrams and photographs. 94pp.
20713-7 Paperbound $1.00

TRICKS AND GAMES ON THE POOL TABLE, Fred Herrmann. 79 tricks and games—some solitaires, some for two or more players, some competitive games—to entertain you between formal games. Mystifying shots and throws, unusual caroms, tricks involving such props as cork, coins, a hat, etc. Formerly *Fun on the Pool Table*. 77 figures. 95pp.
21814-7 Paperbound $1.00

HAND SHADOWS TO BE THROWN UPON THE WALL: A SERIES OF NOVEL AND AMUSING FIGURES FORMED BY THE HAND, Henry Bursill. Delightful picturebook from great-grandfather's day shows how to make 18 different hand shadows: a bird that flies, duck that quacks, dog that wags his tail, camel, goose, deer, boy, turtle, etc. Only book of its sort. vi + 33pp. 6½ x 9¼.
21779-5 Paperbound $1.00

WHITTLING AND WOODCARVING, E. J. Tangerman. 18th printing of best book on market. "If you can cut a potato you can carve" toys and puzzles, chains, chessmen, caricatures, masks, frames, woodcut blocks, surface patterns, much more. Information on tools, woods, techniques. Also goes into serious wood sculpture from Middle Ages to present, East and West. 464 photos, figures. x + 293pp.
20965-2 Paperbound $2.00

HISTORY OF PHILOSOPHY, Julián Marias. Possibly the clearest, most easily followed, best planned, most useful one-volume history of philosophy on the market; neither skimpy nor overfull. Full details on system of every major philosopher and dozens of less important thinkers from pre-Socratics up to Existentialism and later. Strong on many European figures usually omitted. Has gone through dozens of editions in Europe. 1966 edition, translated by Stanley Appelbaum and Clarence Strowbridge. xviii + 505pp.
21739-6 Paperbound $3.00

YOGA: A SCIENTIFIC EVALUATION, Kovoor T. Behanan. Scientific but non-technical study of physiological results of yoga exercises; done under auspices of Yale U. Relations to Indian thought, to psychoanalysis, etc. 16 photos. xxiii + 270pp.
20505-3 Paperbound $2.50

Prices subject to change without notice.
Available at your book dealer or write for free catalogue to Dept. GI, Dover Publications, Inc., 180 Varick St., N. Y., N. Y. 10014. Dover publishes more than 150 books each year on science, elementary and advanced mathematics, biology, music, art, literary history, social sciences and other areas.